ENERGY POLICIES, POLITICS AND PRICES

SOCIAL COST OF CARBON ESTIMATES FOR REGULATORY IMPACT ANALYSIS

DEVELOPMENT AND TECHNICAL ASSESSMENT

ENERGY POLICIES, POLITICS AND PRICES

Additional books in this series can be found on Nova's website under the Series tab.

Additional e-books in this series can be found on Nova's website under the e-book tab.

ENERGY POLICIES, POLITICS AND PRICES

SOCIAL COST OF CARBON ESTIMATES FOR REGULATORY IMPACT ANALYSIS

DEVELOPMENT AND TECHNICAL ASSESSMENT

JOSEPH M. KECK
EDITOR

nova publishers
New York

Library of Congress Cataloging-in-Publication Data

ISBN: 978-1-63463-691-9

Published by Nova Science Publishers, Inc. † New York

CONTENTS

PREFACE

This book presents a summary of the interagency process that developed these SCC estimates.

Chapter 1 – Under Executive Order 12866, agencies are required, to the extent permitted by law, "to assess both the costs and the benefits of the intended regulation and, recognizing that some costs and benefits are difficult to quantify, propose or adopt a regulation only upon a reasoned determination that the benefits of the intended regulation justify its costs." The purpose of the "social cost of carbon" (SCC) estimates presented here is to allow agencies to incorporate the social benefits of reducing carbon dioxide (CO2) emissions into cost-benefit analyses of regulatory actions that have small, or "marginal," impacts on cumulative global emissions. The estimates are presented with an acknowledgement of the many uncertainties involved and with a clear understanding that they should be updated over time to reflect increasing knowledge of the science and economics of climate impacts.

The SCC is an estimate of the monetized damages associated with an incremental increase in carbon emissions in a given year. It is intended to include (but is not limited to) changes in net agricultural productivity, human health, property damages from increased flood risk, and the value of ecosystem services due to climate change.

This document presents a summary of the interagency process that developed these SCC estimates. Technical experts from numerous agencies met on a regular basis to consider public comments, explore the technical literature in relevant fields, and discuss key model inputs and assumptions. The main objective of this process was to develop a range of SCC values using a defensible set of input assumptions grounded in the existing scientific and economic literatures. In this way, key uncertainties and model differences transparently and consistently inform the range of SCC estimates used in the rulemaking process.

The interagency group selected four SCC values for use in regulatory analyses. Three values are based on the average SCC from three integrated assessment models, at discount rates of 2.5, 3, and 5 percent. The fourth value, which represents the 95th percentile SCC estimate across all three models at a 3 percent discount rate, is included to represent higher-than-expected impacts from temperature change further out in the tails of the SCC distribution.

Chapter 2 – Under Executive Order 12866, agencies are required, to the extent permitted by law, "to assess both the costs and the benefits of the intended regulation and, recognizing that some costs and benefits are difficult to quantify, propose or adopt a regulation only upon a reasoned determination that the benefits of the intended regulation justify its costs." The purpose of the "social cost of carbon" (SCC) estimates presented here is to allow agencies to incorporate the social benefits of reducing carbon dioxide (CO2) emissions into cost-benefit analyses of regulatory actions that impact cumulative global emissions. The SCC is an estimate of the monetized damages associated with an incremental increase in carbon emissions in a given year. It is intended to include (but is not limited to) changes in net agricultural productivity, human health, property damages from increased flood risk, and the value of ecosystem services due to climate change.

The interagency process that developed the original U.S. government's SCC estimates is described in the 2010 interagency technical support document (TSD) (Interagency Working Group on Social Cost of Carbon 2010). Through that process the interagency group selected four SCC values for use in regulatory analyses. Three values are based on the average SCC from three integrated assessment models (IAMs), at discount rates of 2.5, 3, and 5 percent. The fourth value, which represents the 95th percentile SCC estimate across all three models at a 3 percent discount rate, is included to represent higher-than-expected impacts from temperature change further out in the tails of the SCC distribution.

While acknowledging the continued limitations of the approach taken by the interagency group in 2010, this document provides an update of the SCC estimates based on new versions of each IAM (DICE, PAGE, and FUND). It does not revisit other interagency modeling decisions (e.g., with regard to the discount rate, reference case socioeconomic and emission scenarios, or equilibrium climate sensitivity). Improvements in the way damages are modeled are confined to those that have been incorporated into the latest versions of the models by the developers themselves in the peer-reviewed literature.

The SCC estimates using the updated versions of the models are higher than those reported in the 2010 TSD. By way of comparison, the four 2020 SCC estimates reported in the 2010 TSD were \$7, \$26, \$42 and \$81 (2007\$). The corresponding four updated SCC estimates for 2020 are \$12, \$43, \$65, and \$129 (2007\$). The model updates that are relevant to the SCC estimates include: an explicit representation of sea level rise damages in the DICE and PAGE models; updated adaptation assumptions, revisions to ensure damages are constrained by GDP, updated regional scaling of damages, and a revised treatment of potentially abrupt shifts in climate damages in the PAGE model; an updated carbon cycle in the DICE model; and updated damage functions for sea level rise impacts, the agricultural sector, and reduced space heating requirements, as well as changes to the transient response of temperature to the buildup of GHG concentrations and the inclusion of indirect effects of methane emissions in the FUND model. The SCC estimates vary by year, and the following table summarizes the revised SCC estimates from 2010 through 2050.

Chapter 3 – Executive Order 12866 directs federal agencies to assess the economic effects of their proposed significant regulatory actions, including a determination that a regulation's benefits justify the costs. In 2008, a federal appeals court directed DOT to update a regulatory impact analysis with an estimate of the social cost of carbon—the dollar value of the net effects (damages and benefits) of an increase in emissions of carbon dioxide, a greenhouse gas.

In 2009, the Interagency Working Group on Social Cost of Carbon was convened to develop estimates for use governmentwide, and it issued final estimates in its 2010 Technical Support Document. In 2013, the group issued revised estimates that were about 50 percent higher than the 2010 estimates, which raised public interest.

GAO was asked to review the working group's development of social cost of carbon estimates. This report describes the participating entities and processes and methods they used to develop the 2010 and 2013 estimates. GAO reviewed executive orders, OMB guidance, the Technical Support Document, its 2013 update, and other key documents. GAO interviewed officials who participated in the working group on behalf of the EOP offices and agencies involved. GAO did not evaluate the quality of the working group's approach.

GAO is making no recommendations in this report. Of seven agencies, OMB and Treasury provided written or oral comments and generally agreed with the findings in this report. Other agencies provided technical comments only or had no comments.

In: Social Cost of Carbon Estimates … ISBN: 978-1-63463-691-9
Editor: Joseph M. Keck

Chapter 1

TECHNICAL SUPPORT DOCUMENT: SOCIAL COST OF CARBON FOR REGULATORY IMPACT ANALYSIS UNDER EXECUTIVE ORDER 12866*

Interagency Working Group on Social Cost of Carbon

EXECUTIVE SUMMARY

Under Executive Order 12866, agencies are required, to the extent permitted by law, "to assess both the costs and the benefits of the intended regulation and, recognizing that some costs and benefits are difficult to quantify, propose or adopt a regulation only upon a reasoned determination that the benefits of the intended regulation justify its costs." The purpose of the "social cost of carbon" (SCC) estimates presented here is to allow agencies to incorporate the social benefits of reducing carbon dioxide (CO_2) emissions into cost-benefit analyses of regulatory actions that have small, or "marginal," impacts on cumulative global emissions. The estimates are presented with an acknowledgement of the many uncertainties involved and with a clear understanding that they should be updated over time to reflect increasing knowledge of the science and economics of climate impacts.

* This is an edited, reformatted and augmented version of a report issued February 2010.

The SCC is an estimate of the monetized damages associated with an incremental increase in carbon emissions in a given year. It is intended to include (but is not limited to) changes in net agricultural productivity, human health, property damages from increased flood risk, and the value of ecosystem services due to climate change.

This document presents a summary of the interagency process that developed these SCC estimates. Technical experts from numerous agencies met on a regular basis to consider public comments, explore the technical literature in relevant fields, and discuss key model inputs and assumptions. The main objective of this process was to develop a range of SCC values using a defensible set of input assumptions grounded in the existing scientific and economic literatures. In this way, key uncertainties and model differences transparently and consistently inform the range of SCC estimates used in the rulemaking process.

The interagency group selected four SCC values for use in regulatory analyses. Three values are based on the average SCC from three integrated assessment models, at discount rates of 2.5, 3, and 5 percent. The fourth value, which represents the 95th percentile SCC estimate across all three models at a 3 percent discount rate, is included to represent higher-than-expected impacts from temperature change further out in the tails of the SCC distribution.

Social Cost of CO_2, 2010 – 2050 (in 2007 dollars)

Discount Rate Year	5% Avg	3% Avg	2.5% Avg	3% 95th
2010	4.7	21.4	35.1	64.9
2015	5.7	23.8	38.4	72.8
2020	6.8	26.3	41.7	80.7
2025	8.2	29.6	45.9	90.4
2030	9.7	32.8	50.0	100.0
2035	11.2	36.0	54.2	109.7
2040	12.7	39.2	58.4	119.3
2045	14.2	42.1	61.7	127.8
2050	15.7	44.9	65.0	136.2

I. Monetizing Carbon Dioxide Emissions

The "social cost of carbon" (SCC) is an estimate of the monetized damages associated with an incremental increase in carbon emissions in a given year. It is intended to include (but is not limited to) changes in net agricultural productivity, human health, property damages from increased flood risk, and the value of ecosystem services. We report estimates of the social cost of carbon in dollars per metric ton of carbon dioxide throughout this document.[1]

When attempting to assess the incremental economic impacts of carbon dioxide emissions, the analyst faces a number of serious challenges. A recent report from the National Academies of Science (NRC 2009) points out that any assessment will suffer from uncertainty, speculation, and lack of information about (1) future emissions of greenhouse gases, (2) the effects of past and future emissions on the climate system, (3) the impact of changes in climate on the physical and biological environment, and (4) the translation of these environmental impacts into economic damages. As a result, any effort to quantify and monetize the harms associated with climate change will raise serious questions of science, economics, and ethics and should be viewed as provisional.

Despite the serious limits of both quantification and monetization, SCC estimates can be useful in estimating the social benefits of reducing carbon dioxide emissions. Under Executive Order 12866, agencies are required, to the extent permitted by law, "to assess both the costs and the benefits of the intended regulation and, recognizing that some costs and benefits are difficult to quantify, propose or adopt a regulation only upon a reasoned determination that the benefits of the intended regulation justify its costs." The purpose of the SCC estimates presented here is to make it possible for agencies to incorporate the social benefits from reducing carbon dioxide emissions into cost-benefit analyses of regulatory actions that have small, or "marginal," impacts on cumulative global emissions. Most federal regulatory actions can be expected to have marginal impacts on global emissions.

For such policies, the benefits from reduced (or costs from increased) emissions in any future year can be estimated by multiplying the change in emissions in that year by the SCC value appropriate for that year. The net present value of the benefits can then be calculated by multiplying each of these future benefits by an appropriate discount factor and summing across all affected years. This approach assumes that the marginal damages from increased emissions are constant for small departures from the baseline emissions path, an approximation that is reasonable for policies that have effects on emissions that are small relative to cumulative global carbon dioxide emissions. For policies that have a large (non-marginal) impact on global cumulative emissions, there is a separate question of whether the SCC is an appropriate tool for calculating the benefits of reduced emissions; we do not attempt to answer that question here.

An interagency group convened on a regular basis to consider public comments, explore the technical literature in relevant fields, and discuss key inputs and assumptions in order to generate SCC estimates. Agencies that actively participated in the interagency process include the Environmental Protection Agency, and the Departments of Agriculture, Commerce, Energy, Transportation, and Treasury. This process was convened by the Council of Economic Advisers and the Office of Management and Budget, with active participation and regular input from the Council on Environmental Quality, National Economic Council, Office of Energy and Climate Change, and Office of Science and Technology Policy. The main objective of this process was to develop a range of SCC values using a defensible set of input assumptions that are grounded in the existing literature. In this way, key uncertainties and model differences can more transparently and consistently inform the range of SCC estimates used in the rulemaking process.

The interagency group selected four SCC estimates for use in regulatory analyses. For 2010, these estimates are $5, $21, $35, and $65 (in 2007 dollars). The first three estimates are based on the average SCC across models and socio-economic and emissions scenarios at the 5, 3, and 2.5 percent discount rates, respectively. The fourth value is included to represent the higher-than-expected impacts from temperature change further out in the tails of the SCC distribution. For this purpose, we use the SCC value for the 95th percentile at a 3 percent discount rate. The central value is the average SCC across models at the 3 percent discount rate. For purposes of capturing the uncertainties involved in regulatory impact analysis, we emphasize the importance and value of considering the full range. These SCC estimates also grow over time. For instance, the central value increases to $24 per ton of CO_2 in 2015 and $26 per ton of CO_2 in 2020. See Appendix A for the full range of annual SCC estimates from 2010 to 2050.

It is important to emphasize that the interagency process is committed to updating these estimates as the science and economic understanding of climate change and its impacts on society improves over time. Specifically, we have set a preliminary goal of revisiting the SCC values within two years or at such time as substantially updated models become available, and to continue to support research in this area. In the meantime, we will continue to explore the issues raised in this document and consider public comments as part of the ongoing interagency process.

II. Social Cost of Carbon Values Used in Past Regulatory Analyses

To date, economic analyses for Federal regulations have used a wide range of values to estimate the benefits associated with reducing carbon dioxide emissions. In the final model year 2011 CAFE rule, the Department of Transportation (DOT) used both a "domestic" SCC value of $2 per ton of CO_2 and a "global" SCC value of $33 per ton of CO_2 for 2007 emission reductions (in 2007 dollars), increasing both values at 2.4 percent per year. It also included a sensitivity analysis at $80 per ton of CO_2. A domestic SCC value is meant to reflect the value of damages in the United States resulting from a unit change in carbon dioxide emissions, while a global SCC value is meant to reflect the value of damages worldwide.

A 2008 regulation proposed by DOT assumed a domestic SCC value of $7 per ton CO_2 (in 2006 dollars) for 2011 emission reductions (with a range of $0-$14 for sensitivity analysis), also increasing at 2.4 percent per year. A regulation finalized by DOE in October of 2008 used a domestic SCC range of $0 to $20 per ton CO_2 for 2007 emission reductions (in 2007 dollars). In addition, EPA's 2008 Advance Notice of Proposed Rulemaking for Greenhouse Gases identified what it described as "very preliminary" SCC estimates subject to revision. EPA's global mean values were $68 and $40 per ton CO_2 for discount rates of approximately 2 percent and 3 percent, respectively (in 2006 dollars for 2007 emissions).

In 2009, an interagency process was initiated to offer a preliminary assessment of how best to quantify the benefits from reducing carbon dioxide emissions. To ensure consistency in how benefits are evaluated across agencies, the Administration sought to develop a transparent and defensible method, specifically designed for the rulemaking process, to quantify avoided climate change damages from reduced CO_2 emissions. The interagency group did not undertake any original analysis. Instead, it combined SCC estimates from the existing literature to use as interim values until a more comprehensive analysis could be conducted.

The outcome of the preliminary assessment by the interagency group was a set of five interim values: global SCC estimates for 2007 (in 2006 dollars) of $55, $33, $19, $10, and $5 per ton of CO_2. The $33 and $5 values represented model-weighted means of the published estimates produced from the most recently available versions of three integrated assessment models—DICE, PAGE, and FUND—at approximately 3 and 5 percent discount rates. The $55 and $10 values were derived by adjusting the published estimates for uncertainty in the discount rate (using factors developed by Newell and Pizer (2003)) at 3 and 5 percent discount rates, respectively. The $19 value was chosen as a central value between the $5 and $33 per ton estimates. All of these values were assumed to increase at 3 percent annually to represent growth in incremental damages over time as the magnitude of climate change increases.

These interim values represent the first sustained interagency effort within the U.S. government to develop an SCC for use in regulatory analysis. The results of this preliminary effort were presented in several proposed and final rules and were offered for public comment in connection with proposed rules, including the joint EPA-DOT fuel economy and CO_2 tailpipe emission proposed rules.

III. Approach and Key Assumptions

Since the release of the interim values, interagency group has reconvened on a regular basis to generate improved SCC estimates. Specifically, the group has considered public comments and further explored the technical literature in relevant fields. This section details the several choices and assumptions that underlie the resulting estimates of the SCC.

It is important to recognize that a number of key uncertainties remain, and that current SCC estimates should be treated as provisional and revisable since they will evolve with improved scientific and economic understanding. The interagency group also recognizes that the existing models are imperfect and incomplete. The National Academy of Science (2009) points out that there is tension between the goal of producing quantified estimates of the economic damages from an incremental ton of carbon and the limits of existing efforts to model these effects. Throughout this document, we highlight a number of concerns and problems that should be addressed by the research community, including research programs housed in many of the agencies participating in the interagency process to estimate the SCC.

The U.S. Government will periodically review and reconsider estimates of the SCC used for cost-benefit analyses to reflect increasing knowledge of the science and economics of climate impacts, as well as improvements in modeling. In this context, statements recognizing the limitations of the analysis and calling for further research take on exceptional significance. The interagency group offers the new SCC values with all due humility about the uncertainties embedded in them and with a sincere promise to continue work to improve them.

A. Integrated Assessment Models

We rely on three integrated assessment models (IAMs) commonly used to estimate the SCC: the FUND, DICE, and PAGE models.[2] These models are frequently cited in the peer-reviewed literature and used in the IPCC assessment. Each model is given equal weight in the SCC values developed through this process, bearing in mind their different limitations (discussed below).

These models are useful because they combine climate processes, economic growth, and feedbacks between the climate and the global economy into a single modeling framework. At the same time, they gain this advantage at the expense of a more detailed representation of the underlying climatic and economic systems. DICE, PAGE, and FUND all take stylized, reduced-form approaches (see NRC 2009 for a more detailed discussion; see Nordhaus 2008 on the possible advantages of this approach). Other IAMs may better reflect the complexity of the science in their modeling frameworks but do not link physical impacts to economic damages. There is currently a limited amount of research linking climate impacts to economic damages, which makes this exercise even more difficult. Underlying the three IAMs selected for this exercise are a number of simplifying assumptions and judgments reflecting the various modelers' best attempts to synthesize the available scientific and economic research characterizing these relationships.

The three IAMs translate emissions into changes in atmospheric greenhouse concentrations, atmospheric concentrations into changes in temperature, and changes in temperature into economic damages. The emissions projections used in the models are based on specified socio-economic (GDP and population) pathways. These emissions are translated into concentrations using the carbon cycle built into each model, and concentrations are translated into warming based on each model's simplified representation of the climate and a key parameter, climate sensitivity. Each model uses a different approach to translate warming into damages. Finally, transforming the stream of economic damages over time into a single value requires judgments about how to discount them.

Each model takes a slightly different approach to model how changes in emissions result in changes in economic damages. In PAGE, for example, the consumption-equivalent damages in each period are calculated as a fraction of GDP, depending on the temperature in that period relative to the preindustrial average temperature in each region. In FUND, damages in each period also depend on the rate of temperature change from the prior period. In DICE, temperature affects both consumption and investment. We describe each model in greater detail here. In a later section, we discuss key gaps in how the models account for various scientific and economic processes (e.g. the probability of catastrophe, and the ability to adapt to climate change and the physical changes it causes).

The parameters and assumptions embedded in the three models vary widely. A key objective of the interagency process was to enable a consistent exploration of the three models while respecting the different approaches to quantifying damages taken by the key modelers in the field. An extensive review of the literature was conducted to select three sets of input parameters for these models: climate sensitivity, socio-economic and emissions trajectories, and discount rates. A probability distribution for climate sensitivity was specified as an input into all three models. In addition, the interagency group used a range of scenarios for the socio-economic parameters and a range of values for the discount rate. All other model features were left unchanged, relying on the model developers' best estimates and judgments. In DICE, these parameters are handled deterministically and represented by fixed constants; in PAGE, most parameters are represented by probability distributions. FUND was also run in a mode in which parameters were treated probabilistically.

The sensitivity of the results to other aspects of the models (e.g. the carbon cycle or damage function) is also important to explore in the context of future revisions to the SCC but has not been incorporated into these estimates. Areas for future research are highlighted at the end of this document.

The DICE Model

The DICE model is an optimal growth model based on a global production function with an extra stock variable (atmospheric carbon dioxide concentrations). Emission reductions are treated as analogous to investment in "natural capital." By investing in natural capital today through reductions in emissions— implying reduced consumption—harmful effects of climate change can be avoided and future consumption thereby increased.

For purposes of estimating the SCC, carbon dioxide emissions are a function of global GDP and the carbon intensity of economic output, with the latter declining over time due to technological progress. The DICE damage function links global average temperature to the overall impact on the world economy. It varies quadratically with temperature change to capture the more rapid increase in damages expected to occur under more extreme climate change, and is calibrated to include the effects of warming on the production of market and nonmarket goods and services. It incorporates impacts on agriculture, coastal areas (due to sea level rise), "other vulnerable market sectors" (based primarily on changes in energy use), human health (based on climate-related diseases, such as malaria and dengue fever, and pollution), non-market amenities (based on outdoor recreation), and human settlements and ecosystems. The DICE damage function also includes the expected value of damages associated with low probability, high impact "catastrophic" climate change. This last component is calibrated based on a survey of experts (Nordhaus 1994). The expected value of these impacts is then added to the other market and non-market impacts mentioned above.

No structural components of the DICE model represent adaptation explicitly, though it is included implicitly through the choice of studies used to calibrate the aggregate damage function. For example, its agricultural impact estimates assume that farmers can adjust land use decisions in response to changing climate conditions, and its health impact estimates assume improvements in healthcare over time. In addition, the small impacts on forestry, water systems, construction, fisheries, and outdoor recreation imply optimistic and costless adaptation in these sectors (Nordhaus and Boyer, 2000; Warren et al., 2006). Costs of resettlement due to sea level rise are incorporated into damage estimates, but their magnitude is not clearly reported. Mastrandrea's (2009) review concludes that "in general, DICE assumes very effective adaptation, and largely ignores adaptation costs."

Note that the damage function in DICE has a somewhat different meaning from the damage functions in FUND and PAGE. Because GDP is endogenous in DICE and because damages in a given year reduce investment in that year, damages propagate forward in time and reduce GDP in future years. In contrast, GDP is exogenous in FUND and PAGE, so damages in any given year do not propagate forward.[3]

The PAGE Model

PAGE2002 (version 1.4epm) treats GDP growth as exogenous. It divides impacts into economic, noneconomic, and catastrophic categories and calculates these impacts separately for eight geographic regions. Damages in each region are expressed as a fraction of output, where the fraction lost depends on the temperature change in each region. Damages are expressed as power functions of temperature change. The exponents of the damage function are the same in all regions but are treated as uncertain, with values ranging from 1 to 3 (instead of being fixed at 2 as in DICE).

PAGE2002 includes the consequences of catastrophic events in a separate damage sub-function. Unlike DICE, PAGE2002 models these events probabilistically. The probability of a "discontinuity" (i.e., a catastrophic event) is assumed to increase with temperature above a specified threshold. The threshold temperature, the rate at which the probability of experiencing a discontinuity increases above the threshold, and the magnitude of the resulting catastrophe are all modeled probabilistically.

Adaptation is explicitly included in PAGE. Impacts are assumed to occur for temperature increases above some tolerable level (2°C for developed countries and 0°C for developing countries for economic impacts, and 0°C for all regions for non-economic impacts), but adaptation is assumed to reduce these impacts. Default values in PAGE2002 assume that the developed countries can ultimately eliminate up to 90 percent of all economic impacts beyond the tolerable 2°C increase and that developing countries can eventually eliminate 50 percent of their economic impacts. All regions are assumed to be able to mitigate 25 percent of the non-economic impacts through adaptation (Hope 2006).

The FUND Model

Like PAGE, the FUND model treats GDP growth as exogenous. It includes separately calibrated damage functions for eight market and nonmarket sectors: agriculture, forestry, water, energy (based on heating and cooling demand), sea level rise (based on the value of land lost and the cost of protection), ecosystems, human health (diarrhea, vector-borne diseases, and cardiovascular and respiratory mortality), and extreme weather. Each impact sector has a different functional form, and is calculated separately for sixteen geographic regions. In some impact sectors, the fraction of output lost or gained due to climate change depends not only on the absolute temperature change but also on the rate of temperature change and level of regional income.[4] In the forestry and agricultural sectors, economic damages also depend on CO_2 concentrations.

Tol (2009) discusses impacts not included in FUND, noting that many are likely to have a relatively small effect on damage estimates (both positive and negative). However, he characterizes several omitted impacts as "big unknowns": for instance, extreme climate scenarios, biodiversity loss, and effects on economic development and political violence. With regard to potentially catastrophic events, he notes, "Exactly what would cause these sorts of changes or what effects they would have are not well-understood, although the chance of any one of them happening seems low. But they do have the potential to happen relatively quickly, and if they did, the costs could be substantial. Only a few studies of climate change have examined these issues."

Adaptation is included both implicitly and explicitly in FUND. Explicit adaptation is seen in the agriculture and sea level rise sectors. Implicit adaptation is included in sectors such as energy and human health, where wealthier populations are assumed to be less vulnerable to climate impacts. For example, the damages to agriculture are the sum of three effects: (1) those due to the rate of temperature change (damages are always positive); (2) those due to the level of temperature change (damages can be positive or negative depending on region and temperature); and (3) those from CO_2 fertilization (damages are generally negative but diminishing to zero).

Adaptation is incorporated into FUND by allowing damages to be smaller if climate change happens more slowly. The combined effect of CO_2 fertilization in the agricultural sector, positive impacts to some regions from higher temperatures, and sufficiently slow increases in temperature across these sectors can result in negative economic damages from climate change.

Damage Functions

To generate revised SCC values, we rely on the IAM modelers' current best judgments of how to represent the effects of climate change (represented by the increase in global-average surface temperature) on the consumption-equivalent value of both market and non-market goods (represented as a fraction of global GDP). We recognize that these representations are incomplete and highly uncertain. But given the paucity of data linking the physical impacts to economic damages, we were not able to identify a better way to translate changes in climate into net economic damages, short of launching our own research program.

The damage functions for the three IAMs are presented in Figures 1A and 1B, using the modeler's default scenarios and mean input assumptions. There are significant differences between the three models both at lower (figure 1B) and higher (figure 1A) increases in global-average temperature.

The lack of agreement among the models at lower temperature increases is underscored by the fact that the damages from FUND are well below the 5th percentile estimated by PAGE, while the damages estimated by DICE are roughly equal to the 95th percentile estimated by PAGE. This is significant because at higher discount rates we expect that a greater proportion of the SCC value is due to damages in years with lower temperature increases. For example, when the discount rate is 2.5 percent, about 45 percent of the 2010 SCC value in DICE is due to damages that occur in years when the temperature is less than or equal to 3 °C. This increases to approximately 55 percent and 80 percent at discount rates of 3 and 5 percent, respectively.

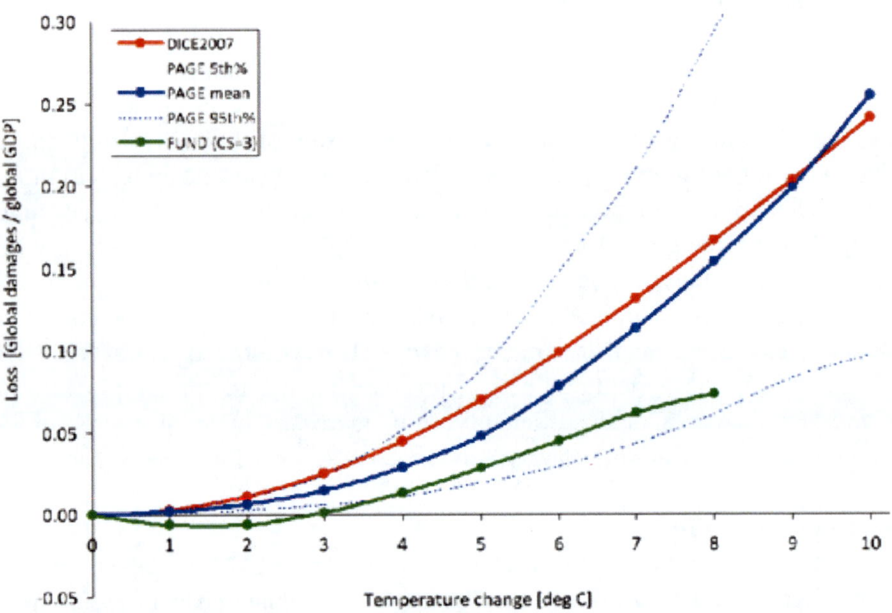

Figure 1A. Annual Consumption Loss as a Fraction of Global GDP in 2100 Due to an Increase in Annual - Global Temperature in the DICE, FUND, and PAGE models.[5]

These differences underscore the need for a thorough review of damage functions—in particular, how the models incorporate adaptation, technological change, and catastrophic damages. Gaps in the literature make modifying these aspects of the models challenging, which highlights the need for additional research. As knowledge improves, the Federal government is committed to exploring how these (and other) models can be modified to incorporate more accurate estimates of damages.

B. Global versus Domestic Measures of SCC

Because of the distinctive nature of the climate change problem, we center our current attention on a global measure of SCC. This approach is the same as that taken for the interim values, but it otherwise represents a departure from past practices, which tended to put greater emphasis on a domestic measure of SCC (limited to impacts of climate change experienced within U.S. borders).

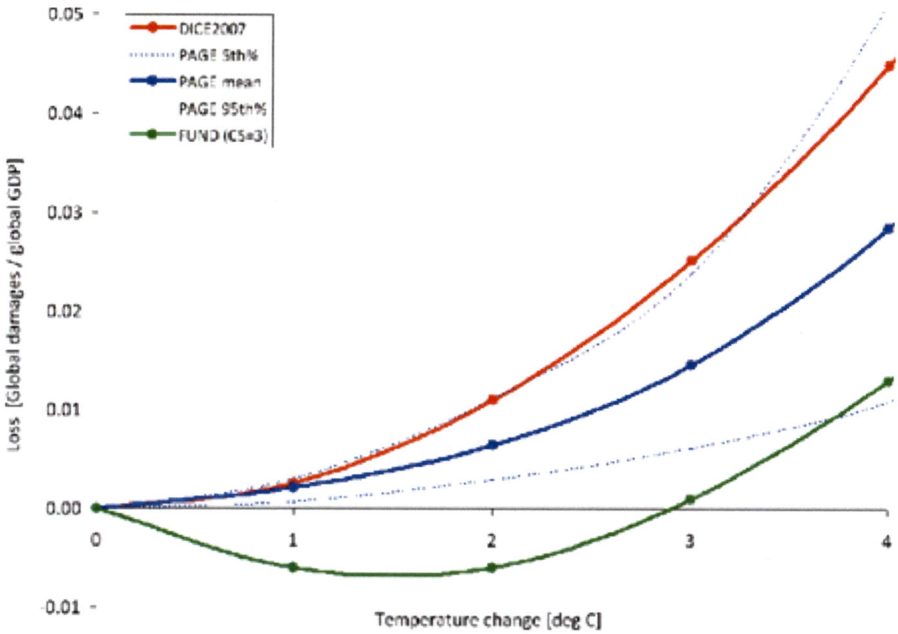

Figure 1B. Annual Consumption Loss for Lower Temperature Changes in DICE, FUND, and PAGE.

As a matter of law, consideration of both global and domestic values is generally permissible; the relevant statutory provisions are usually ambiguous and allow selection of either measure.[6]

Global SCC

Under current OMB guidance contained in Circular A-4, analysis of economically significant proposed and final regulations from the domestic perspective is required, while analysis from the international perspective is optional. However, the climate change problem is highly unusual in at least two respects. First, it involves a global externality: emissions of most greenhouse gases contribute to damages around the world even when they are emitted in the United States. Consequently, to address the global nature of the problem, the SCC must incorporate the full (global) damages caused by GHG emissions. Second, climate change presents a problem that the United States alone cannot solve. Even if the United States were to reduce its greenhouse gas emissions to zero, that step would be far from enough to avoid substantial climate change. Other countries would also need to take action to reduce emissions if significant changes in the global climate are to be avoided.

Emphasizing the need for a global solution to a global problem, the United States has been actively involved in seeking international agreements to reduce emissions and in encouraging other nations, including emerging major economies, to take significant steps to reduce emissions. When these considerations are taken as a whole, the interagency group concluded that a global measure of the benefits from reducing U.S. emissions is preferable.

When quantifying the damages associated with a change in emissions, a number of analysts (e.g., Anthoff, et al. 2009a) employ "equity weighting" to aggregate changes in consumption across regions. This weighting takes into account the relative reductions in wealth in different regions of the world. A per-capita loss of $500 in GDP, for instance, is weighted more heavily in a country with a per-capita GDP of $2,000 than in one with a per-capita GDP of $40,000. The main argument for this approach is that a loss of $500 in a poor country causes a greater reduction in utility or welfare than does the same loss in a wealthy nation. Notwithstanding the theoretical claims on behalf of equity weighting, the interagency group concluded that this approach would not be appropriate for estimating a SCC value used in domestic regulatory analysis.[7] For this reason, the group concluded that using the global (rather than domestic) value, without equity weighting, is the appropriate approach.

Domestic SCC

As an empirical matter, the development of a domestic SCC is greatly complicated by the relatively few region- or country-specific estimates of the SCC in the literature. One potential source of estimates comes from the FUND model. The resulting estimates suggest that the ratio of domestic to global benefits of emission reductions varies with key parameter assumptions. For example, with a 2.5 or 3 percent discount rate, the U.S. benefit is about 7-10 percent of the global benefit, on average, across the scenarios analyzed. Alternatively, if the fraction of GDP lost due to climate change is assumed to be similar across countries, the domestic benefit would be proportional to the U.S. share of global GDP, which is currently about 23 percent.[8]

On the basis of this evidence, the interagency workgroup determined that a range of values from 7 to 23 percent should be used to adjust the global SCC to calculate domestic effects. Reported domestic values should use this range. It is recognized that these values are approximate, provisional, and highly speculative. There is no a priori reason why domestic benefits should be a constant fraction of net global damages over time. Further, FUND does not account for how damages in other regions could affect the United States (e.g., global migration, economic and political destabilization). If more accurate methods for calculating the domestic SCC become available, the Federal government will examine these to determine whether to update its approach.

C. Valuing Non-CO_2 Emissions

While CO_2 is the most prevalent greenhouse gas emitted into the atmosphere, the U.S. included five other greenhouse gases in its recent endangerment finding: methane, nitrous oxide, hydrofluorocarbons, perfluorocarbons, and sulfur hexafluoride. The climate impact of these gases is commonly discussed in terms of their 100-year global warming potential (GWP). GWP measures the ability of different gases to trap heat in the atmosphere (i.e., radiative forcing per unit of mass) over a particular timeframe relative to CO_2. However, because these gases differ in both radiative forcing and atmospheric lifetimes, their relative damages are not constant over time. For example, because methane has a short lifetime, its impacts occur primarily in the near term and thus are not discounted as heavily as those caused by longer-lived gases. Impacts other than temperature change also vary across gases in ways that are not captured by GWP. For instance, CO_2 emissions, unlike methane and other greenhouse gases, contribute to ocean acidification. Likewise, damages from methane emissions are not offset by the positive effect of CO_2 fertilization. Thus, transforming gases into CO_2-equivalents using GWP, and then multiplying the carbon-equivalents by the SCC, would not result in accurate estimates of the social costs of non-CO_2 gases.

In light of these limitations, and the significant contributions of non-CO_2 emissions to climate change, further research is required to link non-CO_2 emissions to economic impacts. Such work would feed into efforts to develop a monetized value of reductions in non-CO_2 greenhouse gas emissions. As part of ongoing work to further improve the SCC estimates, the interagency group hopes to develop methods to value these other greenhouse gases. The goal is to develop these estimates by the time we issue revised SCC estimates for carbon dioxide emissions.

D. Equilibrium Climate Sensitivity

Equilibrium climate sensitivity (ECS) is a key input parameter for the DICE, PAGE, and FUND models.[9] It is defined as the long-term increase in the annual global-average surface temperature from a doubling of atmospheric CO_2 concentration relative to pre-industrial levels (or stabilization at a concentration of approximately 550 parts per million (ppm)). Uncertainties in this important parameter have received substantial attention in the peer-reviewed literature.

The most authoritative statement about equilibrium climate sensitivity appears in the Fourth Assessment Report of the Intergovernmental Panel on Climate Change (IPCC):

> Basing our assessment on a combination of several independent lines of evidence...including observed climate change and the strength of known feedbacks simulated in [global climate models], we conclude that the global mean equilibrium warming for doubling CO_2, or 'equilibrium climate sensitivity', is likely to lie in the range 2 °C to 4.5 °C, with a most likely value of about 3 °C. Equilibrium climate sensitivity is very likely larger than 1.5 °C. [10]
>
> For fundamental physical reasons as well as data limitations, values substantially higher than 4.5 °C still cannot be excluded, but agreement with observations and proxy data is generally worse for those high values than for values in the 2 °C to 4.5 °C range. (Meehl et al., 2007, p 799)

After consulting with several lead authors of this chapter of the IPCC report, the interagency workgroup selected four candidate probability distributions and calibrated them to be consistent with the above statement: Roe and Baker (2007), log-normal, gamma, and Weibull. Table 1 included below gives summary statistics for the four calibrated distributions.

Table 1. Summary Statistics for Four Calibrated Climate Sensitivity Distributions

	Roe & Baker	Log-normal	Gamma	Weibull
Pr(ECS < 1.5°C)	0.013	0.050	0.070	0.102
Pr(2°C < ECS < 4.5°C)	0.667	0.667	0.667	0.667
5th percentile	1.72	1.49	1.37	1.13
10th percentile	1.91	1.74	1.65	1.48
Mode	2.34	2.52	2.65	2.90
Median (50th percentile)	3.00	3.00	3.00	3.00
Mean	3.50	3.28	3.19	3.07
90th percentile	5.86	5.14	4.93	4.69
95th percentile	7.14	5.97	5.59	5.17

Each distribution was calibrated by applying three constraints from the IPCC:

1) a median equal to 3°C, to reflect the judgment of "a most likely value of about 3 °C";[11]
2) two-thirds probability that the equilibrium climate sensitivity lies between 2 and 4.5 °C; and
3) zero probability that it is less than 0°C or greater than 10°C (see Hegerl et al. 2006, p. 721).

We selected the calibrated Roe and Baker distribution from the four candidates for two reasons. First, the Roe and Baker distribution is the only one of the four that is based on a theoretical understanding of the response of the climate system to increased greenhouse gas concentrations (Roe and Baker 2007, Roe 2008). In contrast, the other three distributions are mathematical functions that are arbitrarily chosen based on simplicity, convenience, and general shape. The Roe and Baker distribution results from three assumptions about climate response: (1) absent feedback effects, the equilibrium climate sensitivity is equal to 1.2 °C; (2) feedback factors are proportional to the change in surface temperature; and (3) uncertainties in feedback factors are normally distributed. There is widespread agreement on the first point and the second and third points are common assumptions.

Second, the calibrated Roe and Baker distribution better reflects the IPCC judgment that "values substantially higher than 4.5°C still cannot be excluded." Although the IPCC made no quantitative judgment, the 95th percentile of the calibrated Roe & Baker distribution (7.1 °C) is much closer to the mean and the median (7.2 °C) of the 95th percentiles of 21 previous studies summarized by Newbold and Daigneault (2009). It is also closer to the mean (7.5 °C) and median (7.9 °C) of the nine truncated distributions examined by the IPCC (Hegerl, et al., 2006) than are the 95th percentiles of the three other calibrated distributions (5.2-6.0 °C).

Finally, we note the IPCC judgment that the equilibrium climate sensitivity "is very likely larger than 1.5°C." Although the calibrated Roe & Baker distribution, for which the probability of equilibrium climate sensitivity being greater than 1.5°C is almost 99 percent, is not inconsistent with the IPCC definition of "very likely" as "greater than 90 percent probability," it reflects a greater degree of certainty about very low values of ECS than was expressed by the IPCC.

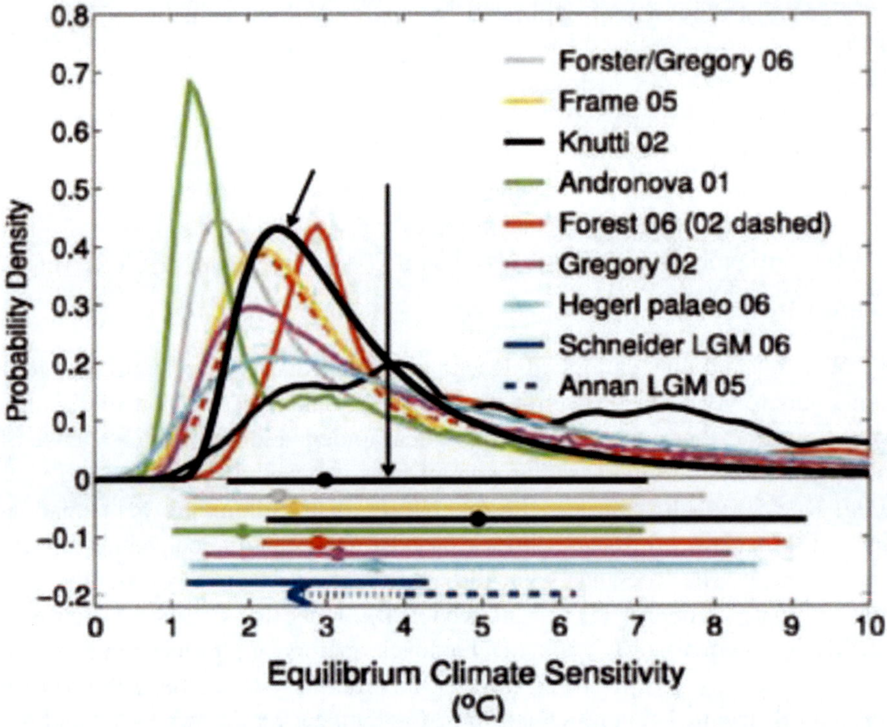

Figure 2. Estimates of the Probability Density Function for Equilibrium Climate Sensitivity (°C).

To show how the calibrated Roe and Baker distribution compares to different estimates of the probability distribution function of equilibrium climate sensitivity in the empirical literature, Figure 2 (below) overlays it on Figure 9.20 from the IPCC Fourth Assessment Report. These functions are scaled to integrate to unity between 0 °C and 10 °C. The horizontal bars show the respective 5 percent to 95 percent ranges; dots indicate the median estimate.[12]

E. Socio-Economic and Emissions Trajectories

Another key issue considered by the interagency group is how to select the set of socio-economic and emissions parameters for use in PAGE, DICE, and FUND. Socio-economic pathways are closely tied to climate damages because, all else equal, more and wealthier people tend to emit more greenhouse gases and also have a higher (absolute) willingness to pay to avoid climate disruptions. For this reason, we consider how to model several input parameters in tandem: GDP, population, CO_2 emissions, and non-CO_2 radiative forcing. A wide variety of scenarios have been developed and used for climate change policy simulations (e.g., SRES 2000, CCSP 2007, EMF 2009). In determining which scenarios are appropriate for inclusion, we aimed to select scenarios that span most of the plausible ranges of outcomes for these variables.

To accomplish this task in a transparent way, we decided to rely on the recent Stanford Energy Modeling Forum exercise, EMF-22. EMF-22 uses ten well-recognized models to evaluate substantial, coordinated global action to meet specific stabilization targets. A key advantage of relying on these data is that GDP, population, and emission trajectories are internally consistent for each model and scenario evaluated. The EMF-22 modeling effort also is preferable to the IPCC SRES due to their age (SRES were developed in 1997) and the fact that 3 of 4 of the SRES scenarios are now extreme outliers in one or more variables. Although the EMF-22 scenarios have not undergone the same level of scrutiny as the SRES scenarios, they are recent, peer-reviewed, published, and publicly available.

To estimate the SCC for use in evaluating domestic policies that will have a small effect on global cumulative emissions, we use socio-economic and emission trajectories that span a range of plausible scenarios. Five trajectories were selected from EMF-22 (see Table 2 below). Four of these represent potential business-as-usual (BAU) growth in population, wealth, and emissions and are associated with CO_2 (only) concentrations ranging from 612 to 889 ppm in 2100. One represents an emissions pathway that achieves stabilization at 550 ppm CO_2e (i.e., CO_2-only concentrations of 425 – 484 ppm or a radiative forcing of 3.7 W/m^2) in 2100, a lower-than-BAU trajectory.[13] Out of the 10 models included in the EMF-22 exercise, we selected the trajectories used by MiniCAM, MESSAGE, IMAGE, and the optimistic scenario from MERGE. For the BAU pathways, we used the GDP, population, and emission trajectories from each of these four models. For the 550 ppm CO_2e scenario, we averaged the GDP, population, and emission trajectories implied by these same four models.

We explore how sensitive the SCC is to various assumptions about how the future will evolve without prejudging what is likely to occur. The interagency group considered formally assigning probability weights to different states of the world, but this proved challenging to do in an analytically rigorous way given the dearth of information on the likelihood of a full range of future socio-economic pathways.

Table 2. Socioeconomic and Emissions Projections from Select EMF-22 Reference Scenarios -Reference Fossil and Industrial CO_2 Emissions ($GtCO_2$/yr)

EMF – 22 Based Scenarios	2000	2010	2020	2030	2050	2100
IMAGE	26.6	31.9	36.9	40.0	45.3	60.1
MERGE Optimistic	24.6	31.5	37.6	45.1	66.5	117.9
MESSAGE	26.8	29.2	37.6	42.1	43.5	42.7
MiniCAM	26.5	31.8	38.0	45.1	57.8	80.5
550 ppm average	26.2	31.1	33.2	32.4	20.0	12.8
	Reference GDP (using market exchange rates in trillion 2005$)[14]					
EMF – 22 Based Scenarios	2000	2010	2020	2030	2050	2100
IMAGE	38.6	53.0	73.5	97.2	156.3	396.6
MERGE Optimistic	36.3	45.9	59.7	76.8	122.7	268.0
MESSAGE	38.1	52.3	69.4	91.4	153.7	334.9
MiniCAM	36.1	47.4	60.8	78.9	125.7	369.5
550 ppm average	37.1	49.6	65.6	85.5	137.4	337.9
EMF – 22 Based Scenarios	Global Population (billions)					
	2000	2010	2020	2030	2050	2100
IMAGE	6.1	6.9	7.6	8.2	9.0	9.1
MERGE Optimistic	6.0	6.8	7.5	8.2	9.0	9.7
MESSAGE	6.1	6.9	7.7	8.4	9.4	10.4
MiniCAM	6.0	6.8	7.5	8.1	8.8	8.7
550 ppm average	6.1	6.8	7.6	8.2	8.7	9.1

There are a number of caveats. First, EMF BAU scenarios represent the modelers' judgment of the most likely pathway absent mitigation policies to reduce greenhouse gas emissions, rather than the wider range of possible outcomes. Nevertheless, these views of the most likely outcome span a wide range, from the more optimistic (e.g. abundant low-cost, low-carbon energy) to more pessimistic (e.g. constraints on the availability of nuclear and renewables).[15] Second, the socio-economic trajectories associated with a 550 ppm CO_2e concentration scenario are not derived from an assessment of what policy is optimal from a benefit-cost standpoint. Rather, it is indicative of one possible future outcome. The emission trajectories underlying some BAU scenarios (e.g. MESSAGE's 612 ppm) also are consistent with some modest policy action to address climate change.[16] We chose not to include socio-economic trajectories that achieve even lower GHG concentrations at this time, given the difficulty many models had in converging to meet these targets.

For comparison purposes, the Energy Information Agency in its 2009 Annual Energy Outlook projected that global carbon dioxide emissions will grow to 30.8, 35.6, and 40.4 gigatons in 2010, 2020, and 2030, respectively, while world GDP is projected to be $51.8, $71.0 and $93.9 trillion (in 2005 dollars using market exchange rates) in 2010, 2020, and 2030, respectively. These projections are consistent with one or more EMF-22 scenarios. Likewise, the United Nations' 2008 Population Prospect projects population will grow from 6.1 billion people in 2000 to 9.1 billion people in 2050, which is close to the population trajectories for the IMAGE, MiniCAM, and MERGE models.

In addition to fossil and industrial CO_2 emissions, each EMF scenario provides projections of methane, nitrous oxide, fluorinated greenhouse gases, and net land use CO_2 emissions out to 2100. These assumptions also are used in the three models while retaining the default radiative forcings due to other factors (e.g. aerosols and other gases). See the Appendix for greater detail.

F. Discount Rate

The choice of a discount rate, especially over long periods of time, raises highly contested and exceedingly difficult questions of science, economics, philosophy, and law. Although it is well understood that the discount rate has a large influence on the current value of future damages, there is no consensus about what rates to use in this context. Because carbon dioxide emissions are long-lived, subsequent damages occur over many years. In calculating the SCC, we first estimate the future damages to agriculture, human health, and other market and non-market sectors from an additional unit of carbon dioxide emitted in a particular year in terms of reduced consumption (or consumption equivalents) due to the impacts of elevated temperatures, as represented in each of the three IAMs. Then we discount the stream of future damages to its present value in the year when the additional unit of emissions was released using the selected discount rate, which is intended to reflect society's marginal rate of substitution between consumption in different time periods.

For rules with both intra- and intergenerational effects, agencies traditionally employ constant discount rates of both 3 percent and 7 percent in accordance with OMB Circular A-4. As Circular A-4 acknowledges, however, the choice of discount rate for intergenerational problems raises distinctive problems and presents considerable challenges. After reviewing those challenges, Circular A-4 states, "If your rule will have important intergenerational benefits or costs you might consider a further sensitivity analysis using a lower but positive discount rate in addition to calculating net benefits using discount rates of 3 and 7 percent." For the specific purpose of developing the SCC, we adapt and revise that approach here.

Arrow et al. (1996) outlined two main approaches to determine the discount rate for climate change analysis, which they labeled "descriptive" and "prescriptive." The descriptive approach reflects a positive (non-normative) perspective based on observations of people's actual choices—e.g., savings versus consumption decisions over time, and allocations of savings among more and less risky investments. Advocates of this approach generally call for inferring the discount rate from market rates of return "because of a lack of justification for choosing a social welfare function that is any different than what decision makers [individuals] actually use" (Arrow et al. 1996).

One theoretical foundation for the cost-benefit analyses in which the social cost of carbon will be used— the Kaldor-Hicks potential-compensation test— also suggests that market rates should be used to discount future benefits and costs, because it is the market interest rate that would govern the returns potentially set aside today to compensate future individuals for climate damages that they bear (e.g., Just et al. 2004). As some have noted, the word "potentially" is an important qualification; there is no assurance that such returns will actually be set aside to provide compensation, and the very idea of compensation is difficult to define in the intergenerational context. On the other hand, societies provide compensation to future generations through investments in human capital and the resulting increase in knowledge, as well as infrastructure and other physical capital.

The prescriptive approach specifies a social welfare function that formalizes the normative judgments that the decision-maker wants explicitly to incorporate into the policy evaluation—e.g., how interpersonal comparisons of utility should be made, and how the welfare of future generations should be weighed against that of the present generation. Ramsey (1928), for example, has argued that it is "ethically indefensible" to apply a positive pure rate of time preference to discount values across generations, and many agree with this view.

Other concerns also motivate making adjustments to descriptive discount rates. In particular, it has been noted that the preferences of future generations with regard to consumption versus environmental amenities may not be the same as those today, making the current market rate on consumption an inappropriate metric by which to discount future climate-related damages.

Others argue that the discount rate should be below market rates to correct for market distortions and uncertainties or inefficiencies in intergenerational transfers of wealth, which in the Kaldor-Hicks logic are presumed to compensate future generations for damage (a potentially controversial assumption, as noted above) (Arrow et al. 1996, Weitzman 1999).

Further, a legitimate concern about both descriptive and prescriptive approaches is that they tend to obscure important heterogeneity in the population. The utility function that underlies the prescriptive approach assumes a representative agent with perfect foresight and no credit constraints. This is an artificial rendering of the real world that misses many of the frictions that characterize individuals' lives and indeed the available descriptive evidence supports this. For instance, many individuals smooth consumption by borrowing with credit cards that have relatively high rates. Some are unable to access traditional credit markets and rely on payday lending operations or other high cost forms of smoothing consumption. Whether one puts greater weight on the prescriptive or descriptive approach, the high interest rates that credit-constrained individuals accept suggest that some account should be given to the discount rates revealed by their behavior.

We draw on both approaches but rely primarily on the descriptive approach to inform the choice of discount rate. With recognition of its limitations, we find this approach to be the most defensible and transparent given its consistency with the standard contemporary theoretical foundations of benefit-cost analysis and with the approach required by OMB's existing guidance. The logic of this framework also suggests that market rates should be used for discounting future consumption-equivalent damages. Regardless of the theoretical approach used to derive the appropriate discount rate(s), we note the inherent conceptual and practical difficulties of adequately capturing consumption trade-offs over many decades or even centuries. While relying primarily on the descriptive approach in selecting specific discount rates, the interagency group has been keenly aware of the deeply normative dimensions of both the debate over discounting in the intergenerational context and the consequences of selecting one discount rate over another.

Historically Observed Interest Rates

In a market with no distortions, the return to savings would equal the private return on investment, and the market rate of interest would be the appropriate choice for the social discount rate. In the real world risk, taxes, and other market imperfections drive a wedge between the risk-free rate of return on capital and the consumption rate of interest. Thus, the literature recognizes two conceptual discount concepts—the consumption rate of interest and the opportunity cost of capital.

According to OMB's Circular A-4, it is appropriate to use the rate of return on capital when a regulation is expected to displace or alter the use of capital in the private sector. In this case, OMB recommends Agencies use a discount rate of 7 percent. When regulation is expected to primarily affect private consumption—for instance, via higher prices for goods and services—a lower discount rate of 3 percent is appropriate to reflect how private individuals trade-off current and future consumption.

The interagency group examined the economics literature and concluded that the consumption rate of interest is the correct concept to use in evaluating the benefits and costs of a marginal change in carbon emissions (see Lind 1990, Arrow et al 1996, and Arrow 2000). The consumption rate of interest also is appropriate when the impacts of a regulation are measured in consumption (-equivalent) units, as is done in the three integrated assessment models used for estimating the SCC.

Individuals use a variety of savings instruments that vary with risk level, time horizon, and tax characteristics. The standard analytic framework used to develop intuition about the discount rate typically assumes a representative agent with perfect foresight and no credit constraints. The risk-free rate is appropriate for discounting certain future benefits or costs, but the benefits calculated by IAMs are uncertain. To use the risk-free rate to discount uncertain benefits, these benefits first must be transformed into "certainty equivalents," that is the maximum certain amount that we would exchange for the uncertain amount. However, the calculation of the certainty-equivalent requires first estimating the correlation between the benefits of the policy and baseline consumption.

If the IAM projections of future impacts represent expected values (not certainty-equivalent values), then the appropriate discount rate generally does not equal the risk-free rate. If the benefits of the policy tend to be high in those states of the world in which consumption is low, then the certainty-equivalent benefits will be higher than the expected benefits (and vice versa). Since many (though not necessarily all) of the important impacts of climate change will flow through market sectors such as agriculture and energy, and since willingness to pay for environmental protections typically increases with income, we might expect a positive (though not necessarily perfect) correlation between the net benefits from climate policies and market returns. This line of reasoning suggests that the proper discount rate would exceed the riskless rate. Alternatively, a negative correlation between the returns to climate policies and market returns would imply that a discount rate below the riskless rate is appropriate.

This discussion suggests that both the post-tax riskless and risky rates can be used to capture individuals' consumption-equivalent interest rate. As a measure of the post-tax riskless rate, we calculate the average real return from Treasury notes over the longest time period available (those from Newell and Pizer 2003) and adjust for Federal taxes (the average marginal rate from tax years 2003 through 2006 is around 27 percent).[17] This calculation produces a real interest rate of about 2.7 percent, which is roughly consistent with Circular A-4's recommendation to use 3 percent to represent the consumption rate of interest.[18] A measure of the post-tax risky rate for investments whose returns are positively correlated with overall equity market returns can be obtained by adjusting pre-tax rates of household returns to risky investments (approximately 7 percent) for taxes yields a real rate of roughly 5 percent.[19]

The Ramsey Equation

Ramsey discounting also provides a useful framework to inform the choice of a discount rate. Under this approach, the analyst applies either positive or normative judgments in selecting values for the key parameters of the Ramsey equation: ç (coefficient of relative risk aversion or elasticity of the marginal utility of consumption) and p (pure rate of time preference).[20] These are then combined with g (growth rate of per-capita consumption) to equal the interest rate at which future monetized damages are discounted: $\rho + \eta \cdot g$.[21] In the simplest version of the Ramsey model, with an optimizing representative agent with perfect foresight, what we are calling the "Ramsey discount rate," $\rho + \eta \cdot g$, will be equal to the rate of return to capital, i.e., the market interest rate.

A review of the literature provides some guidance on reasonable parameter values for the Ramsey discounting equation, based on both prescriptive and descriptive approaches.

- η. Most papers in the climate change literature adopt values for ç in the range of 0.5 to 3 (Weitzman cites plausible values as those ranging from 1 to 4), although not all authors articulate whether their choice is based on prescriptive or descriptive reasoning.[22] Dasgupta (2008) argues that η should be greater than 1 and may be as high as 3, since η equal to 1 suggests savings rates that do not conform to observed behavior.
- p. With respect to the pure rate of time preference, most papers in the climate change literature adopt values for ρ in the range of 0 to 3

percent per year. The very low rates tend to follow from moral judgments involving intergenerational neutrality. Some have argued that to use any value other than p = 0 would unjustly discriminate against future generations (e.g., Arrow et al. 1996, Stern et al. 2006). However, even in an inter-generational setting, it may make sense to use a small positive pure rate of time preference because of the small probability of unforeseen cataclysmic events (Stern et al. 2006).

- g. A commonly accepted approximation is around 2 percent per year. For the socio-economic scenarios used for this exercise, the EMF models assume that g is about 1.5-2 percent to 2100.

Some economists and non-economists have argued for constant discount rates below 2 percent based on the prescriptive approach. When grounded in the Ramsey framework, proponents of this approach have argued that a p of zero avoids giving preferential treatment to one generation over another. The choice of ç has also been posed as an ethical choice linked to the value of an additional dollar in poorer countries compared to wealthier ones. Stern et al. (2006) applies this perspective through his choice of p = 0.1 percent per year, 11 = 1 and g = 1.3 percent per year, which yields an annual discount rate of 1.4 percent. In the context of permanent income savings behavior, however, Stern's assumptions suggest that individuals would save 93 percent of their income.[23]

Recently, Stern (2008) revisited the values used in Stern et al. (2006), stating that there is a case to be made for raising 11 due to the amount of weight lower values place on damages far in the future (over 90 percent of expected damages occur after 2200 with 11 = 1). Using Stern's assumption that p = 0.1 percent, combined with a 11 of 1.5 to 2 and his original growth rate, yields a discount rate greater 2 percent.

We conclude that arguments made under the prescriptive approach can be used to justify discount rates between roughly 1.4 and 3.1 percent. In light of concerns about the most appropriate value for 11, we find it difficult to justify rates at the lower end of this range under the Ramsey framework.

Accounting for Uncertainty in the Discount Rate

While the consumption rate of interest is an important driver of the benefits estimate, it is uncertain over time. Ideally, we would formally model this uncertainty, just as we do for climate sensitivity. Weitzman (1998, 2001) showed theoretically and Newell and Pizer (2003) and Groom et al. (2006) confirm empirically that discount rate uncertainty can have a large effect on net present values. A main result from these studies is that if there is a persistent element to the uncertainty in the discount rate (e.g., the rate follows a random walk), then it will result in an effective (or certainty-equivalent) discount rate that declines over time. Consequently, lower discount rates tend to dominate over the very long term (see Weitzman 1998, 1999, 2001; Newell and Pizer 2003; Groom et al. 2006; Gollier 2008; Summers and Zeckhauser 2008; and Gollier and Weitzman 2009).

The proper way to model discount rate uncertainty remains an active area of research. Newell and Pizer (2003) employ a model of how long-term interest rates change over time to forecast future discount rates. Their model incorporates some of the basic features of how interest rates move over time, and its parameters are estimated based on historical observations of long-term rates. Subsequent work on this topic, most notably Groom et al. (2006), uses more general models of interest rate dynamics to allow for better forecasts. Specifically, the volatility of interest rates depends on whether rates are currently low or high and variation in the level of persistence over time.

While Newell and Pizer (2003) and Groom et al (2006) attempt formally to model uncertainty in the discount rate, others argue for a declining scale of discount rates applied over time (e.g., Weitzman 2001, and the UK's "Green Book" for regulatory analysis). This approach uses a higher discount rate initially, but applies a graduated scale of lower discount rates further out in time.[24] A key question that has emerged with regard to both of these approaches is the trade-off between potential time inconsistency and giving greater weight to far future outcomes (see the EPA Science Advisory Board's recent comments on this topic as part of its review of their *Guidelines for Economic Analysis*).[25]

The Discount Rates Selected for Estimating SCC

In light of disagreement in the literature on the appropriate market interest rate to use in this context and uncertainty about how interest rates may change over time, we use three discount rates to span a plausible range of certainty-equivalent constant discount rates: 2.5, 3, and 5 percent per year. Based on the review in the previous sections, the interagency workgroup determined that these three rates reflect reasonable judgments under both descriptive and prescriptive approaches.

The central value, 3 percent, is consistent with estimates provided in the economics literature and OMB's Circular A-4 guidance for the consumption rate of interest. As previously mentioned, the consumption rate of interest is the correct discounting concept to use when future damages from elevated temperatures are estimated in consumption-equivalent units. Further, 3 percent roughly corresponds to the after-tax riskless interest rate. The upper value of 5 percent is included to represent the possibility that climate damages are positively correlated with market returns. Additionally, this discount rate may be justified by the high interest rates that many consumers use to smooth consumption across periods.

The low value, 2.5 percent, is included to incorporate the concern that interest rates are highly uncertain over time. It represents the average certainty-equivalent rate using the mean-reverting and random walk approaches from Newell and Pizer (2003) starting at a discount rate of 3 percent. Using this approach, the certainty equivalent is about 2.2 percent using the random walk model and 2.8 percent using the mean reverting approach.[26] Without giving preference to a particular model, the average of the two rates is 2.5 percent. Further, a rate below the riskless rate would be justified if climate investments are negatively correlated with the overall market rate of return. Use of this lower value also responds to certain judgments using the prescriptive or normative approach and to ethical objections that have been raised about rates of 3 percent or higher.

IV. Revised SCC Estimates

Our general approach to estimating SCC values is to run the three integrated assessment models (FUND, DICE, and PAGE) using the following inputs agreed upon by the interagency group:

- A Roe and Baker distribution for the climate sensitivity parameter bounded between 0 and 10 with a median of 3 °C and a cumulative probability between 2 and 4.5 °C of two-thirds.
- Five sets of GDP, population and carbon emissions trajectories based on EMF-22.
- Constant annual discount rates of 2.5, 3, and 5 percent.

Because the climate sensitivity parameter is modeled probabilistically, and because PAGE and FUND incorporate uncertainty in other model parameters, the final output from each model run is a distribution over the SCC in year *t*.

For each of the IAMS, the basic computational steps for calculating the SCC in a particular year *t* are:

1) Input the path of emissions, GDP, and population from the selected EMF-22 scenarios, and the extrapolations based on these scenarios for post-2100 years.
2) Calculate the temperature effects and (consumption-equivalent) damages in each year resulting from the baseline path of emissions.
   a) In PAGE, the consumption-equivalent damages in each period are calculated as a fraction of the EMF GDP forecast, depending on the temperature in that period relative to the pre-industrial average temperature in each region.
   b) In FUND, damages in each period depend on both the level and the rate of temperature change in that period.
   c) In DICE, temperature affects both consumption and investment, so we first adjust the EMF GDP paths as follows: Using the Cobb-Douglas production function with the DICE2007 parameters, we extract the path of exogenous technical change implied by the EMF GDP and population paths, then we recalculate the baseline GDP path taking into account climate damages resulting from the baseline emissions path.
3) Add an additional unit of carbon emissions in year t. (The exact unit varies by model.)
4) Recalculate the temperature effects and damages expected in all years beyond tresulting from this adjusted path of emissions, as in step 2.
5) Subtract the damages computed in step 2 from those in step 4 in each year. (DICE is run in 10 year time steps, FUND in annual time steps, while the time steps in PAGE vary.)
6) Discount the resulting path of marginal damages back to the year of emissions using the agreed upon fixed discount rates.
7) Calculate the SCC as the net present value of the discounted path of damages computed in step 6, divided by the unit of carbon emissions used to shock the models in step 3.
8) Multiply by 12/44 to convert from dollars per ton of carbon to dollars per ton of CO_2 (2007 dollars) in DICE and FUND. (All calculations are done in tons of CO2 in PAGE).

The steps above were repeated in each model for multiple future years to cover the time horizons anticipated for upcoming rulemaking analysis. To maintain consistency across the three IAMs, climate damages are calculated as lost consumption in each future year.

It is important to note that each of the three models has a different default end year. The default time horizon is 2200 for PAGE, 2595 for DICE, and 3000 for the latest version of FUND. This is an issue for the multi-model approach because differences in SCC estimates may arise simply due to the model time horizon. Many consider 2200 too short a time horizon because it could miss a significant fraction of damages under certain assumptions about the growth of marginal damages and discounting, so each model is run here through 2300. This step required a small adjustment in the PAGE model only. This step also required assumptions about GDP, population, and greenhouse gas emission trajectories after 2100, the last year for which these data are available from the EMF-22 models. (A more detailed discussion of these assumptions is included in the Appendix.)

This exercise produces 45 separate distributions of the SCC for a given year, the product of 3 models, 3 discount rates, and 5 socioeconomic scenarios. This is clearly too many separate distributions for consideration in a regulatory impact analysis.

To produce a range of plausible estimates that still reflects the uncertainty in the estimation exercise, the distributions from each of the models and scenarios are equally weighed and combined to produce three separate probability distributions for SCC in a given year, one for each assumed discount rate. These distributions are then used to define a range of point estimates for the global SCC. In this way, no integrated assessment model or socioeconomic scenario is given greater weight than another. Because the literature shows that the SCC is quite sensitive to assumptions about the discount rate, and because no consensus exists on the appropriate rate to use in an intergenerational context, we present SCCs based on the average values across models and socioeconomic scenarios for each discount rate.

The interagency group selected four SCC values for use in regulatory analyses. Three values are based on the average SCC across models and socio-economic and emissions scenarios at the 2.5, 3, and 5 percent discount rates. The fourth value is included to represent the higher-than-expected economic impacts from climate change further out in the tails of the SCC distribution. For this purpose, we use the SCC value for the 95th percentile at a 3 percent discount rate. (The full set of distributions by model and scenario combination is included in the Appendix.) As noted above, the 3 percent discount rate is the central value, and so the central value that emerges is the average SCC across models at the 3 percent discount rate. For purposes of capturing the uncertainties involved in regulatory impact analysis, we emphasize the importance and value of considering the full range.

As previously discussed, low probability, high impact events are incorporated into the SCC values through explicit consideration of their effects in two of the three models as well as the use of a probability density function for equilibrium climate sensitivity. Treating climate sensitivity probabilistically results in more high temperature outcomes, which in turn lead to higher projections of damages. Although FUND does not include catastrophic damages (in contrast to the other two models), its probabilistic treatment of the equilibrium climate sensitivity parameter will directly affect the non-catastrophic damages that are a function of the rate of temperature change.

In Table 3, we begin by presenting SCC estimates for 2010 by model, scenario, and discount rate to illustrate the variability in the SCC across each of these input parameters. As expected, higher discount rates consistently result in lower SCC values, while lower discount rates result in higher SCC values for each socioeconomic trajectory. It is also evident that there are differences in the SCC estimated across the three main models. For these estimates, FUND produces the lowest estimates, while PAGE generally produces the highest estimates.

Table 3. Disaggregated Social Cost of CO_2 Values by Model, Socio-Economic Trajectory, and Discount Rate for 2010 (in 2007 dollars)

Model	Discount rate: Scenario	5% Avg	3% Avg	2.5% Avg	3% 95th
DICE	IMAGE	10.8	35.8	54.2	70.8
	MERGE	7.5	22.0	31.6	42.1
	Message	9.8	29.8	43.5	58.6
	MiniCAM	8.6	28.8	44.4	57.9
	550 Average	8.2	24.9	37.4	50.8
PAGE	IMAGE	8.3	39.5	65.5	142.4
	MERGE	5.2	22.3	34.6	82.4
	Message	7.2	30.3	49.2	115.6
	MiniCAM	6.4	31.8	54.7	115.4
	550 Average	5.5	25.4	42.9	104.7
FUND	IMAGE	-1.3	8.2	19.3	39.7
	MERGE	-0.3	8.0	14.8	41.3
	Message	-1.9	3.6	8.8	32.1
	MiniCAM	-0.6	10.2	22.2	42.6
	550 Average	-2.7	-0.2	3.0	19.4

These results are not surprising when compared to the estimates in the literature for the latest versions of each model. For example, adjusting the values from the literature that were used to develop interim SCC values to 2007 dollars for the year 2010 (assuming, as we did for the interim process, that SCC grows at 3 percent per year), FUND yields SCC estimates at or near zero for a 5 percent discount rate and around $9 per ton for a 3 percent discount rate. There are far fewer estimates using the latest versions of DICE and PAGE in the literature: Using similar adjustments to generate 2010 estimates, we calculate a SCC from DICE (based on Nordhaus 2008) of around $9 per ton for a 5 percent discount rate, and a SCC from PAGE (based on Hope 2006, 2008) close to $8 per ton for a 4 percent discount rate. Note that these comparisons are only approximate since the literature generally relies on Ramsey discounting, while we have assumed constant discount rates.[27]

The SCC estimates from FUND are sensitive to differences in emissions paths but relatively insensitive to differences in GDP paths across scenarios, while the reverse is true for DICE and PAGE. This likely occurs because of several structural differences among the models. Specifically in DICE and PAGE, the fraction of economic output lost due to climate damages increases with the level of temperature alone, whereas in FUND the fractional loss also increases with the rate of temperature change. Furthermore, in FUND increases in income over time decrease vulnerability to climate change (a form of adaptation), whereas this does not occur in DICE and PAGE. These structural differences among the models make FUND more sensitive to the path of emissions and less sensitive to GDP compared to DICE and PAGE.

Figure 3 shows that IMAGE has the highest GDP in 2100 while MERGE Optimistic has the lowest. The ordering of global GDP levels in 2100 directly corresponds to the rank ordering of SCC for PAGE and DICE. For FUND, the correspondence is less clear, a result that is to be expected given its less direct relationship between its damage function and GDP.

Table 4 shows the four selected SCC values in five year increments from 2010 to 2050. Values for 2010, 2020, 2040, and 2050 are calculated by first combining all outputs (10,000 estimates per model run) from all scenarios and models for a given discount rate. Values for the years in between are calculated using a simple linear interpolation.

The SCC increases over time because future emissions are expected to produce larger incremental damages as physical and economic systems become more stressed in response to greater climatic change. Note that this approach allows us to estimate the growth rate of the SCC directly using DICE, PAGE, and FUND rather than assuming a constant annual growth rate as was done for the interim estimates (using 3 percent).

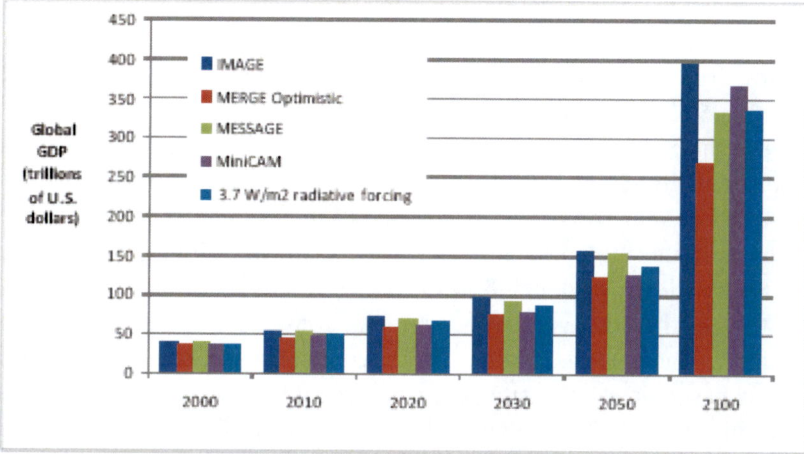

Figure 3. Level of Global GDP across EMF Scenarios.

Table 4. Social Cost of CO_2, 2010 – 2050 (in 2007 dollars)

Discount Rate Year	5% Avg	3% Avg	2.5% Avg	3% 95th
2010	4.7	21.4	35.1	64.9
2015	5.7	23.8	38.4	72.8
2020	6.8	26.3	41.7	80.7
2025	8.2	29.6	45.9	90.4
2030	9.7	32.8	50.0	100.0
2035	11.2	36.0	54.2	109.7
2040	12.7	39.2	58.4	119.3
2045	14.2	42.1	61.7	127.8
2050	15.7	44.9	65.0	136.2

Table 5. Changes in the Average Annual Growth Rates of SCC Estimates between 2010 and 2050

Average Annual Growth Rate (%)	**5%** Avg	**3%** Avg	**2.5%** Avg	**3.0%** 95th
2010-2020	3.6%	2.1%	1.7%	2.2%
2020-2030	3.7%	2.2%	1.8%	2.2%
2030-2040	2.7%	1.8%	1.6%	1.8%
2040-2050	2.1%	1.4%	1.1%	1.3%

This helps to ensure that the estimates are internally consistent with other modeling assumptions. Table 5 illustrates how the growth rate for these four SCC estimates varies over time. The full set of annual SCC estimates between 2010 and 2050 is reported in the Appendix.

While the SCC estimate grows over time, the future monetized value of emissions reductions in each year (the SCC in year *t* multiplied by the change in emissions in year *t*) must be discounted to the present to determine its total net present value for use in regulatory analysis. Damages from future emissions should be discounted at the same rate as that used to calculate the SCC estimates themselves to ensure internal consistency—i.e., future damages from climate change, whether they result from emissions today or emissions in a later year, should be discounted using the same rate. For example, climate damages in the year 2020 that are calculated using a SCC based on a 5 percent discount rate also should be discounted back to the analysis year using a 5 percent discount rate.[28]

V. LIMITATIONS OF THE ANALYSIS

As noted, any estimate of the SCC must be taken as provisional and subject to further refinement (and possibly significant change) in accordance with evolving scientific, economic, and ethical understandings. During the course of our modeling, it became apparent that there are several areas in particular need of additional exploration and research. These caveats, and additional observations in the following section, are necessary to consider when interpreting and applying the SCC estimates.

Incomplete treatment of non-catastrophic damages. The impacts of climate change are expected to be widespread, diverse, and heterogeneous. In addition, the exact magnitude of these impacts is uncertain because of the inherent complexity of climate processes, the economic behavior of current and future populations, and our inability to accurately forecast technological change and adaptation. Current IAMs do not assign value to all of the important physical, ecological, and economic impacts of climate change recognized in the climate change literature (some of which are discussed above) because of lack of precise information on the nature of damages and because the science incorporated into these models understandably lags behind the most recent research. Our ability to quantify and monetize impacts will undoubtedly improve with time. But it is also likely that even in future applications, a number of potentially significant damage categories will remain non-monetized. (Ocean acidification is one example of a potentially large damage from CO_2 emissions not quantified by any of the three models. Species and wildlife loss is another example that is exceedingly difficult to monetize.)

Incomplete treatment of potential catastrophic damages. There has been considerable recent discussion of the risk of catastrophic impacts and how best to account for extreme scenarios, such as the collapse of the Atlantic Meridional Overturning Circulation or the West Antarctic Ice Sheet, or large releases of methane from melting permafrost and warming oceans. Weitzman (2009) suggests that catastrophic damages are extremely large—so large, in fact, that the damages from a low probability, catastrophic event far in the future dominate the effect of the discount rate in a present value calculation and result in an infinite willingness-to-pay for mitigation today. However, Nordhaus (2009) concluded that the conditions under which Weitzman's results hold "are limited and do not apply to a wide range of potential uncertain scenarios."

Using a simplified IAM, Newbold and Daigneault (2009) confirmed the potential for large catastrophe risk premiums but also showed that the aggregate benefit estimates can be highly sensitive to the shapes of both the climate sensitivity distribution and the damage function at high temperature changes. Pindyck (2009) also used a simplified IAM to examine high-impact low-probability risks, using a right-skewed gamma distribution for climate sensitivity as well as an uncertain damage coefficient, but in most cases found only a modest risk premium. Given this difference in opinion, further research in this area is needed before its practical significance can be fully understood and a reasonable approach developed to account for such risks in regulatory analysis. (The next section discusses the scientific evidence on catastrophic impacts in greater detail.)

Uncertainty in extrapolation of damages to high temperatures: The damage functions in these IAMs are typically calibrated by estimating damages at moderate temperature increases (e.g., DICE was calibrated at 2.5 °C) and extrapolated to far higher temperatures by assuming that damages increase as some power of the temperature change. Hence, estimated damages are far more uncertain under more extreme climate change scenarios.

Incomplete treatment of adaptation and technological change: Each of the three integrated assessment models used here assumes a certain degree of low- or no-cost adaptation. For instance, Tol assumes a great deal of adaptation in FUND, including widespread reliance on air conditioning ; so much so, that the largest single benefit category in FUND is the reduced electricity costs from not having to run air conditioning as intensively (NRC 2009).

Climate change also will increase returns on investment to develop technologies that allow individuals to cope with adverse climate conditions, and IAMs to do not adequately account for this directed technological change.[29] For example, scientists may develop crops that are better able to withstand higher and more variable temperatures. Although DICE and FUND have both calibrated their agricultural sectors under the assumption that farmers will change land use practices in response to climate change (Mastrandrea, 2009), they do not take into account technological changes that lower the cost of this adaptation over time. On the other hand, the calibrations do not account for increases in climate variability, pests, or diseases, which could make adaptation more difficult than assumed by the IAMs for a given temperature change. Hence, models do not adequately account for potential adaptation or technical change that might alter the emissions pathway and resulting damages. In this respect, it is difficult to determine whether the incomplete treatment of adaptation and technological change in these IAMs under or overstate the likely damages.

Risk aversion: A key question unanswered during this interagency process is what to assume about relative risk aversion with regard to high-impact outcomes. These calculations do not take into account the possibility that individuals may have a higher willingness to pay to reduce the likelihood of low-probability, high-impact damages than they do to reduce the likelihood of higher-probability but lower-impact damages with the same expected cost. (The inclusion of the 95th percentile estimate in the final set of SCC values was largely motivated by this concern.) If individuals do show such a higher willingness to pay, a further question is whether that fact should be taken into account for regulatory policy. Even if individuals are not risk-averse for such scenarios, it is possible that regulatory policy should include a degree of risk-aversion.

Assuming a risk-neutral representative agent is consistent with OMB's Circular A-4, which advises that the estimates of benefits and costs used in regulatory analysis are usually based on the average or the expected value and that "emphasis on these expected values is appropriate as long as society is 'risk neutral' with respect to the regulatory alternatives. While this may not always be the case, [analysts] should in general assume 'risk neutrality' in [their] analysis."

Nordhaus (2008) points to the need to explore the relationship between risk and income in the context of climate change across models and to explore the role of uncertainty regarding various parameters in the results. Using FUND, Anthoff et al (2009) explored the sensitivity of the SCC to Ramsey equation parameter assumptions based on observed behavior. They conclude that “the assumed rate of risk aversion is at least as important as the assumed rate of time preference in determining the social cost of carbon.” Since Circular A-4 allows for a different assumption on risk preference in regulatory analysis if it is adequately justified, we plan to continue investigating this issue.

VI. A FURTHER DISCUSSION OF CATASTROPHIC IMPACTS AND DAMAGE FUNCTIONS

As noted above, the damage functions underlying the three IAMs used to estimate the SCC may not capture the economic effects of all possible adverse consequences of climate change and may therefore lead to underestimates of the SCC (Mastrandrea 2009). In particular, the models’ functional forms may not adequately capture: (1) potentially discontinuous “tipping point” behavior in Earth systems, (2) inter-sectoral and inter-regional interactions, including global security impacts of high-end warming, and (3) limited near-term substitutability between damage to natural systems and increased consumption.

It is the hope of the interagency group that over time researchers and modelers will work to fill these gaps and that the SCC estimates used for regulatory analysis by the Federal government will continue to evolve with improvements in modeling. In the meantime, we discuss some of the available evidence.

*Extrapolation of climate damages to high levels of warming.*The damage functions in the models are calibrated at moderate levels of warming and should therefore be viewed cautiously when extrapolated to the high temperatures found in the upper end of the distribution. Recent science suggests that there are a number of potential climatic “tipping points” at which the Earth system may exhibit discontinuous behavior with potentially severe social and economic consequences (e.g., Lenton et al, 2008, Kriegler et al., 2009). These tipping points include the disruption of the Indian Summer Monsoon, dieback of the Amazon Rainforest and boreal forests, collapse of the Greenland Ice Sheet and the West Antarctic Ice Sheet, reorganization of the Atlantic Meridional Overturning Circulation, strengthening of El Niño-Southern Oscillation, and the release of methane from melting permafrost. Many of these tipping points are estimated to have thresholds between about 3

°C and 5 °C (Lenton et al., 2008). Probabilities of several of these tipping points were assessed through expert elicitation in 2005–2006 by Kriegler et al. (2009); results from this study are highlighted in Table 6. Ranges of probability are averaged across core experts on each topic.

As previously mentioned, FUND does not include potentially catastrophic effects. DICE assumes a small probability of catastrophic damages that increases with increased warming, but the damages from these risks are incorporated as expected values (i.e., ignoring potential risk aversion). PAGE models catastrophic impacts in a probabilistic framework (see Figure 1), so the high-end output from PAGE potentially offers the best insight into the SCC if the world were to experience catastrophic climate change. For instance, at the 95th percentile and a 3 percent discount rate, the SCC estimated by PAGE across the five socio-economic and emission trajectories of \$113 per ton of CO_2 is almost double the value estimated by DICE, \$58 per ton in 2010. We cannot evaluate how well the three models account for catastrophic or non-catastrophic impacts, but this estimate highlights the sensitivity of SCC values in the tails of the distribution to the assumptions made about catastrophic impacts.

Table 6. Probabilities of Various Tipping Points from Expert Elicitation

Possible Tipping Points	Duration before effect is fully realized (in years)	Additional Warming by 2100		
		0.5-1.5 C	1.5-3.0 C	3-5 C
Reorganization of Atlantic Meridional Overturning Circulation	about 100	0-18%	6-39%	18-67%
Greenland Ice Sheet collapse	at least 300	8-39%	33-73%	67-96%
West Antarctic Ice Sheet collapse	at least 300	5-41%	10-63%	33-88%
Dieback of Amazon rainforest	about 50	2-46%	14-84%	41-94%
Strengthening of El Niño-Southern Oscillation	about 100	1-13%	6-32%	19-49%
Dieback of boreal forests	about 50	13-43%	20-81%	34-91%
Shift in Indian Summer Monsoon	about 1	Not formally assessed		
Release of methane from melting permafrost	Less than 100	Not formally assessed.		

PAGE treats the possibility of a catastrophic event probabilistically, while DICE treats it deterministically (that is, by adding the expected value of the damage from a catastrophe to the aggregate damage function). In part, this results in different probabilities being assigned to a catastrophic event across the two models. For instance, PAGE places a probability near zero on a catastrophe at 2.5 °C warming, while DICE assumes a 4 percent probability of a catastrophe at 2.5 °C. By comparison, Kriegler et al. (2009) estimate a probability of at least 16-36 percent of crossing at least one of their primary climatic tipping points in a scenario with temperatures about 2-4 °C warmer than pre-Industrial levels in 2100.

It is important to note that crossing a climatic tipping point will not necessarily lead to an economic catastrophe in the sense used in the IAMs. A tipping point is a critical threshold across which some aspect of the Earth system starts to shifts into a qualitatively different state (for instance, one with dramatically reduced ice sheet volumes and higher sea levels). In the IAMs, a catastrophe is a low-probability environmental change with high economic impact.

Failure to incorporate inter-sectoral and inter-regional interactions. The damage functions do not fully incorporate either inter-sectoral or inter-regional interactions. For instance, while damages to the agricultural sector are incorporated, the effects of changes in food supply on human health are not fully captured and depend on the modeler's choice of studies used to calibrate the IAM. Likewise, the effects of climate damages in one region of the world on another region are not included in some of the models (FUND includes the effects of migration from sea level rise). These inter-regional interactions, though difficult to quantify, are the basis for climate-induced national and economic security concerns (e.g., Campbell et al., 2007; U.S. Department of Defense 2010) and are particularly worrisome at higher levels of warming. High-end warming scenarios, for instance, project water scarcity affecting 4.3-6.9 billion people by 2050, food scarcity affecting about 120 million additional people by 2080, and the creation of millions of climate refugees (Easterling et al., 2007; Campbell et al., 2007).

Imperfect substitutability of environmental amenities. Data from the geological record of past climate changes suggests that 6 °C of warming may have severe consequences for natural systems. For instance, during the Paleocene-Eocene Thermal Maximum about 55.5 million years ago, when the Earth experienced a geologically rapid release of carbon associated with an approximately 5 °C increase in global mean temperatures, the effects included shifts of about 400-900 miles in the range of plants (Wing et al., 2005), and dwarfing of both land mammals (Gingerich, 2006) and soil fauna (Smith et al., 2009).

The three IAMs used here assume that it is possible to compensate for the economic consequences of damages to natural systems through increased consumption of non-climate goods, a common assumption in many economic models. In the context of climate change, however, it is possible that the damages to natural systems could become so great that no increase in consumption of non-climate goods would provide complete compensation (Levy et al., 2005). For instance, as water supplies become scarcer or ecosystems become more fragile and less bio-diverse, the services they provide may become increasingly more costly to replace. Uncalibrated attempts to incorporate the imperfect substitutability of such amenities into IAMs (Sterner and Persson, 2008) indicate that the optimal degree of emissions abatement can be considerably greater than is commonly recognized.

CONCLUSION

The interagency group selected four SCC estimates for use in regulatory analyses. For 2010, these estimates are \$5, \$21, \$35, and \$65 (in 2007 dollars). The first three estimates are based on the average SCC across models and socio-economic and emissions scenarios at the 5, 3, and 2.5 percent discount rates, respectively. The fourth value is included to represent the higher-than-expected impacts from temperature change further out in the tails of the SCC distribution. For this purpose, we use the SCC value for the 95th percentile at a 3 percent discount rate. The central value is the average SCC across models at the 3 percent discount rate. For purposes of capturing the uncertainties involved in regulatory impact analysis, we emphasize the importance and value of considering the full range. These SCC estimates also grow over time. For instance, the central value increases to \$24 per ton of CO_2 in 2015 and \$26 per ton of CO_2 in 2020.

We noted a number of limitations to this analysis, including the incomplete way in which the integrated assessment models capture catastrophic and non-catastrophic impacts, their incomplete treatment of adaptation and technological change, uncertainty in the extrapolation of damages to high temperatures, and assumptions regarding risk aversion. The limited amount of research linking climate impacts to economic damages makes this modeling exercise even more difficult. It is the hope of the interagency group that over time researchers and modelers will work to fill these gaps and that the SCC estimates used for regulatory analysis by the Federal government will continue to evolve with improvements in modeling.

APPENDIX

Table A1. Annual SCC Values: 2010–2050 (in 2007 dollars)

Discount Rate Year	5% Avg	3% Avg	2.5% Avg	3% 95th
2010	4.7	21.4	35.1	64.9
2011	4.9	21.9	35.7	66.5
2012	5.1	22.4	36.4	68.1
2013	5.3	22.8	37.0	69.6
2014	5.5	23.3	37.7	71.2
2015	5.7	23.8	38.4	72.8
2016	5.9	24.3	39.0	74.4
2017	6.1	24.8	39.7	76.0
2018	6.3	25.3	40.4	77.5
2019	6.5	25.8	41.0	79.1
2020	6.8	26.3	41.7	80.7
2021	7.1	27.0	42.5	82.6
2022	7.4	27.6	43.4	84.6
2023	7.7	28.3	44.2	86.5
2024	7.9	28.9	45.0	88.4
2025	8.2	29.6	45.9	90.4
2026	8.5	30.2	46.7	92.3
2027	8.8	30.9	47.5	94.2
2028	9.1	31.5	48.4	96.2
2029	9.4	32.1	49.2	98.1
2030	9.7	32.8	50.0	100.0
2031	10.0	33.4	50.9	102.0
2032	10.3	34.1	51.7	103.9
2033	10.6	34.7	52.5	105.8
2034	10.9	35.4	53.4	107.8
2035	11.2	36.0	54.2	109.7
2036	11.5	36.7	55.0	111.6
2037	11.8	37.3	55.9	113.6
2038	12.1	37.9	56.7	115.5
2039	12.4	38.6	57.5	117.4
2040	12.7	39.2	58.4	119.3
2041	13.0	39.8	59.0	121.0
2042	13.3	40.4	59.7	122.7
2043	13.6	40.9	60.4	124.4
2044	13.9	41.5	61.0	126.1
2045	14.2	42.1	61.7	127.8
2046	14.5	42.6	62.4	129.4

Table A1. (Continued)

Discount Rate Year	5% Avg	3% Avg	2.5% Avg	3% 95th
2047	14.8	43.2	63.0	131.1
2048	15.1	43.8	63.7	132.8
2049	15.4	44.4	64.4	134.5
2050	15.7	44.9	65.0	136.2-39

This Appendix also provides additional technical information about the non-CO_2 emission projections used in the modeling and the method for extrapolating emissions forecasts through 2300, and shows the full distribution of 2010 SCC estimates by model and scenario combination.

1. Other (non-CO_2) gases

In addition to fossil and industrial CO_2 emissions, each EMF scenario provides projections of methane (CH_4), nitrous oxide (N_2O), fluorinated gases, and net land use CO_2 emissions to 2100. These assumptions are used in all three IAMs while retaining each model's default radiative forcings (RF) due to other factors (e.g., aerosols and other gases). Specifically, to obtain the RF associated with the nonCO_2 EMF emissions only, we calculated the RF associated with the EMF atmospheric CO_2 concentrations and subtracted them from the EMF total RF.[30] This approach respects the EMF scenarios as much as possible and at the same time takes account of those components not included in the EMF projections. Since each model treats non-CO_2 gases differently (e.g., DICE lumps all other gases into one composite exogenous input), this approach was applied slightly differently in each of the models.

FUND: Rather than relying on RF for these gases, the actual emissions from each scenario were used in FUND. The model default trajectories for CH_4, N_20, SF_6, and the CO_2 emissions from land were replaced with the EMF values.

PAGE: PAGE models CO_2, CH_4, sulfur hexafluoride (SF_6), and aerosols and contains an "excess forcing" vector that includes the RF for everything else. To include the EMF values, we removed the default CH_4 and SF_6 factors[31], decomposed the excess forcing vector, and constructed a new excess forcing vector that includes the EMF RF for CH_4, N_20, and fluorinated gases, as well as the model default values for aerosols and other factors. Net land use CO_2 emissions were added to the fossil and industrial CO_2 emissions pathway.

DICE: DICE presents the greatest challenge because all forcing due to factors other than industrial CO_2 emissions is embedded in an exogenous non-CO_2 RF vector. To decompose this exogenous forcing path into EMF non-CO_2 gases and other gases, we relied on the references in DICE2007 to the Intergovernmental Panel on Climate Change's (IPCC) Fourth Assessment Report (AR4) and the discussion of aerosol forecasts in the IPCC's Third Assessment Report (TAR) and in AR4, as explained below. In DICE2007, Nordhaus assumes that exogenous forcing from all non-CO_2 sources is -0.06 W/m^2 in 2005, as reported in AR4, and increases linearly to 0.3 W/m^2 in 2105, based on GISS projections, and then stays constant after that time.

According to AR4, the RF in 2005 from CH_4, N_20, and halocarbons (approximately similar to the F-gases in the EMF-22 scenarios) was 0.48 + 0.16 + 0.34 = 0.98 W/m^2 and RF from total aerosols was -1.2 W/m^2. Thus, the -.06 W/m^2 non-CO_2 forcing in DICE can be decomposed into: 0.98 W/m^2 due to the EMF nonCO_2 gases, -1.2 W/m^2 due to aerosols, and the remainder, 0.16 W/m^2, due to other residual forcing.

For subsequent years, we calculated the DICE default RF from aerosols and other non-CO_2 gases based on the following two assumptions:

1) RF from aerosols declines linearly from 2005 to 2100 at the rate projected by the TAR and then stays constant thereafter, and
2) With respect to RF from non-CO_2 gases not included in the EMF-22 scenarios, the share of non-aerosol RF matches the share implicit in the AR4 summary statistics cited above and remains constant over time.

Assumption (1) means that the RF from aerosols in 2100 equals 66 percent of that in 2000, which is the fraction of the TAR projection of total RF from aerosols (including sulfates, black carbon, and organic carbon) in 2100 vs. 2000 under the A1B SRES emissions scenario. Since the SRES marker scenarios were not updated for the AR4, the TAR provides the most recent IPCC projection of aerosol forcing. We rely on the A1B projection from the TAR because it provides one of the lower aerosol forecasts among the SRES marker scenarios and is more consistent with the AR4 discussion of the post-SRES literature on aerosols:

Aerosols have a net cooling effect and the representation of aerosol and aerosol precursor emissions, including sulphur dioxide, black carbon and organic carbon, has improved in the post-SRES scenarios. Generally, these emissions are projected to be lower than reported in SRES. {WGIII 3.2, TS.3, SPM}.[32]

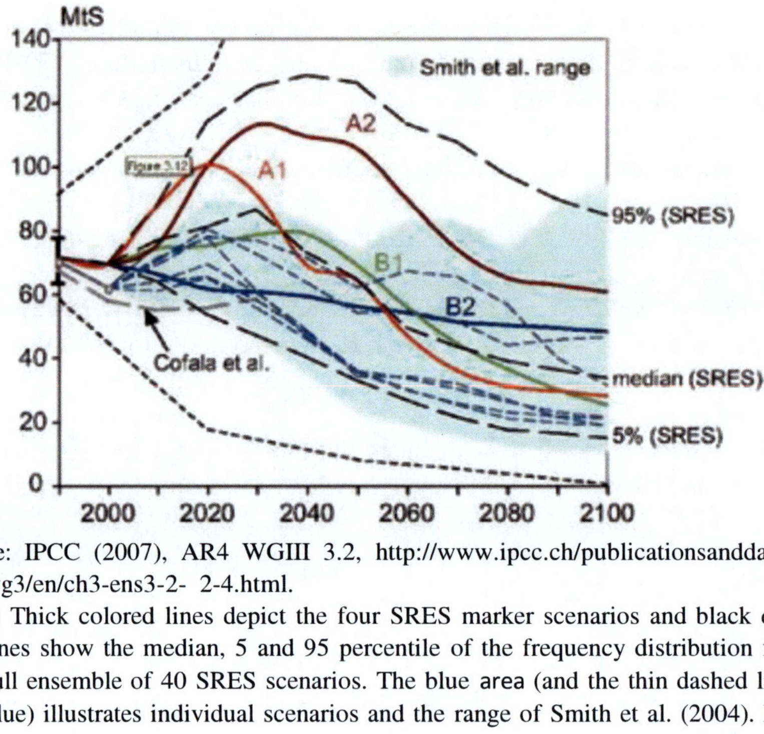

Source: IPCC (2007), AR4 WGIII 3.2, http://www.ipcc.ch/publicationsanddata/ar4/wg3/en/ch3-ens3-2- 2-4.html.

Notes: Thick colored lines depict the four SRES marker scenarios and black dashed lines show the median, 5 and 95 percentile of the frequency distribution for the full ensemble of 40 SRES scenarios. The blue area (and the thin dashed lines in blue) illustrates individual scenarios and the range of Smith et al. (2004). Dotted lines indicate the minimum and maximum of SO2 emissions scenarios developed pre-SRES.

Figure A1. Sulphur Dioxide Emission Scenarios.

Assuming a simple linear decline in aerosols from 2000 to 2100 also is more consistent with the recent literature on these emissions. For example, Figure A1 shows that the sulfur dioxide emissions peak over the short-term of some SRES scenarios above the upper bound estimates of the more recent scenarios.[33] Recent scenarios project sulfur emissions to peak earlier and at lower levels compared to the SRES in part because of new information about present and planned sulfur legislation in some developing countries, such as

India and China.[34] The lower bound projections of the recent literature have also shifted downward slightly compared to the SRES scenario (IPCC 2007).

With these assumptions, the DICE aerosol forcing changes from -1.2 in 2005 to -0.792 in 2105 W/m^2; forcing due to other non-CO_2 gases not included in the EMF scenarios declines from 0.160 to 0.153 W/m^2.

Although other approaches to decomposing the DICE exogenous forcing vector are possible, initial sensitivity analysis suggests that the differences among reasonable alternative approaches are likely to be minor. For example, adjusting the TAR aerosol projection above to assume that aerosols will be maintained at 2000 levels through 2100 reduces average SCC values (for 2010) by approximately 3 percent (or less than $2); assuming all aerosols are phased out by 2100 increases average 2010 SCC values by 6-7 percent (or $0.50-$3)–depending on the discount rate. These differences increase slightly for SCC values in later years but are still well within 10 percent of each other as far out as 2050.

Finally, as in PAGE, the EMF net land use CO_2 emissions are added to the fossil and industrial CO_2 emissions pathway.

2. Extrapolating Emissions Projections to 2300

To run each model through 2300 requires assumptions about GDP, population, greenhouse gas emissions, and radiative forcing trajectories after 2100, the last year for which these projections are available from the EMF-22 models. These inputs were extrapolated from 2100 to 2300 as follows:

1) Population growth rate declines linearly, reaching zero in the year 2200.
2) GDP/ per capita growth rate declines linearly, reaching zero in the year 2300.
3) The decline in the fossil and industrial carbon intensity (CO_2/GDP) growth rate over 2090-2100 is maintained from 2100 through 2300.
4) Net land use CO_2 emissions decline linearly, reaching zero in the year 2200.
5) Non-CO_2 radiative forcing remains constant after 2100.

Long run stabilization of GDP per capita was viewed as a more realistic simplifying assumption than a linear or exponential extrapolation of the pre-

2100 economic growth rate of each EMF scenario. This is based on the idea that increasing scarcity of natural resources and the degradation of environmental sinks available for assimilating pollution from economic production activities may eventually overtake the rate of technological progress. Thus, the overall rate of economic growth may slow over the very long run. The interagency group also considered allowing an exponential decline in the growth rate of GDP per capita. However, since this would require an additional assumption about how close to zero the growth rate would get by 2300, the group opted for the simpler and more transparent linear extrapolation to zero by 2300.

The population growth rate is also assumed to decline linearly, reaching zero by 2200. This assumption is reasonably consistent with the United Nations long run population forecast, which estimates global population to be fairly stable after 2150 in the medium scenario (UN 2004).[35] The resulting range of EMF population trajectories (Figure A2) also encompass the UN medium scenario forecasts through 2300 – global population of 8.5 billion by 2200, and 9 billion by 2300.

Maintaining the decline in the 2090-2100 carbon intensity growth rate (i.e., CO_2 per dollar of GDP) through 2300 assumes that technological improvements and innovations in the areas of energy efficiency and other carbon reducing technologies (possibly including currently unavailable methods) will continue to proceed at roughly the same pace that is projected to occur towards the end of the forecast period for each EMF scenario. This assumption implies that total cumulative emissions in 2300 will be between 5,000 and 12,000 GtC, which is within the range of the total potential global carbon stock estimated in the literature.

Net land use CO_2 emissions are expected to stabilize in the long run, so in the absence of any post 2100 projections, the group assumed a linear decline to zero by 2200. Given no a priori reasons for assuming a long run increase or decline in non-CO_2 radiative forcing, it is assumed to remain at the 2100 levels for each EMF scenario through 2300.

Figures A2-A7 show the paths of global population, GDP, fossil and industrial CO_2 emissions, net land CO_2 emissions, non-CO_2 radiative forcing, and CO_2 intensity (fossil and industrial CO_2 emissions/GDP) resulting from these assumptions.

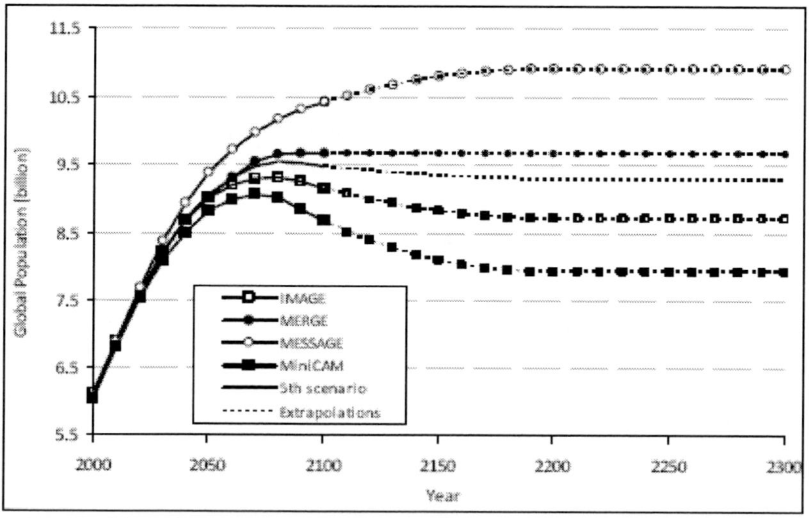

Note: In the fifth scenario, 2000-2100 population is equal to the average of the population under the 550 ppm CO2e, full-participation, not-to-exceed scenarios considered by each of the four models.

Figure A2. Global Population, 2000-2300 (Post-2100 extrapolations assume the population growth.

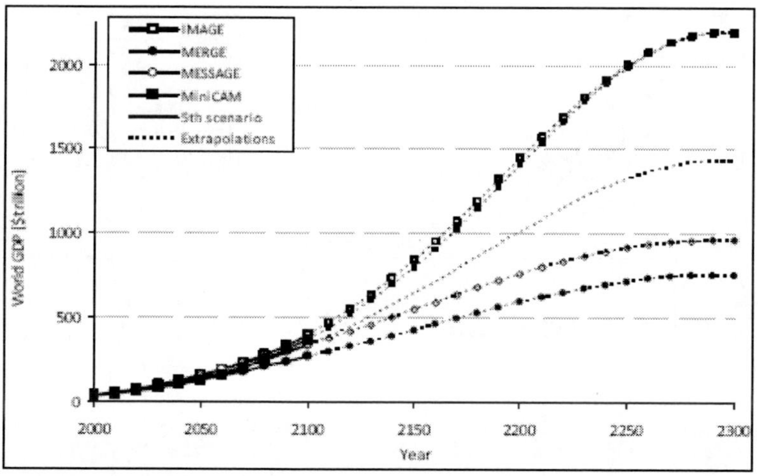

Note: In the fifth scenario, 2000-2100 GDP is equal to the average of the GDP under the 550 ppm CO2e, full-participation, not-to-exceed scenarios considered by each of the four models.

Figure A3. World GDP, 2000-2300 (Post-2100 extrapolations assume GDP per capita growth declines linearly, reaching zero in the year 2300).

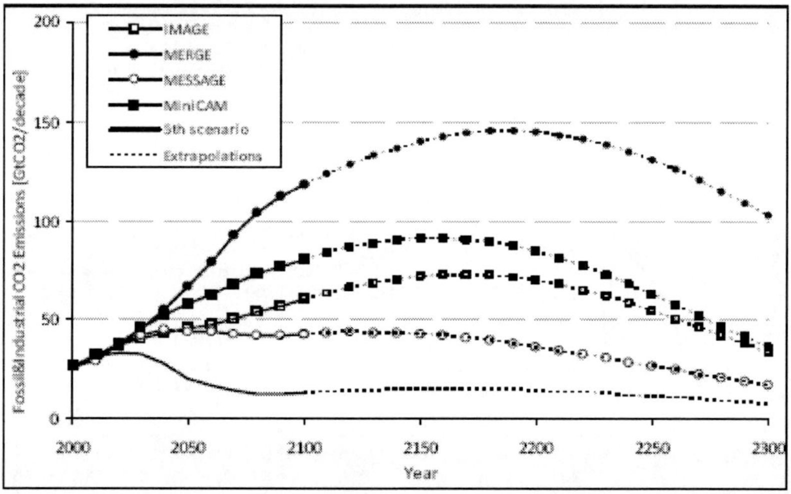

Note: In the fifth scenario, 2000-2100 emissions are equal to the average of the emissions under the 550 ppm CO2e, full-participation, not-to-exceed scenarios considered by each of the four models.

Figure A4. Global Fossil and Industrial CO_2 Emissions, 2000-2300 (Post-2100 extrapolations assume growth rate of CO_2 intensity (CO_2/GDP) over 2090-2100 is maintained through 2300).

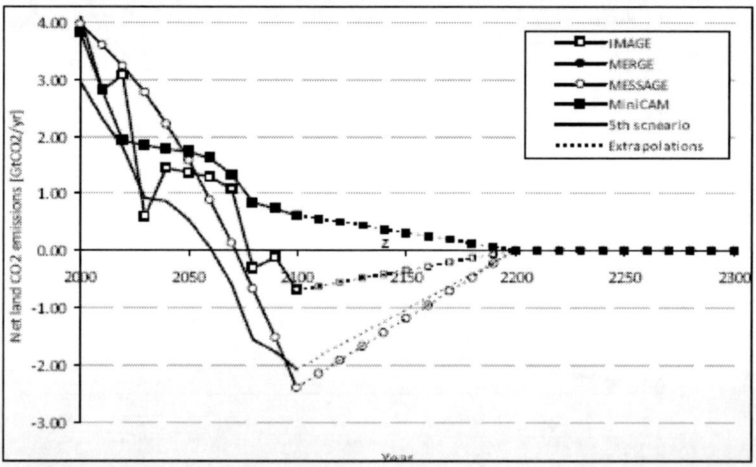

Note: In the fifth scenario, 2000-2100 emissions are equal to the average of the emissions under the 550 ppm CO2e, full-participation, not-to-exceed scenarios considered by each of the four models.

Figure A5. Global Net Land Use CO_2 Emissions, 2000-2300 (Post-2100 extrapolations assume emissions decline linearly, reaching zero in the year 2200).[36]

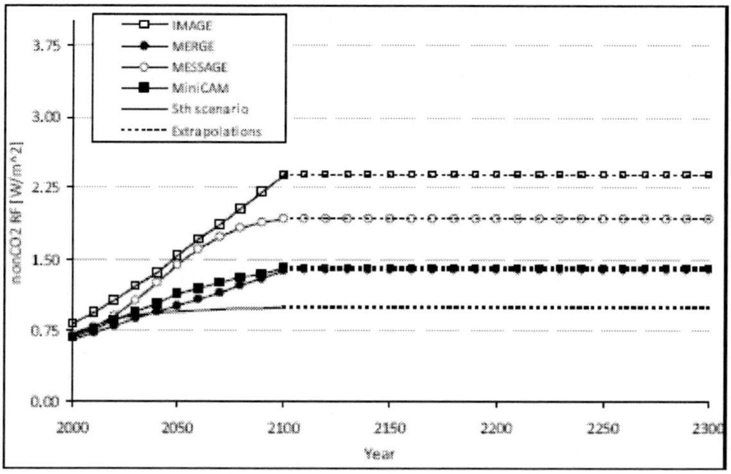

Note: In the fifth scenario, 2000-2100 emissions are equal to the average of the emissions under the 550 ppm CO2e, full-participation, not-to-exceed scenarios considered by each of the four models.

Figure A6. Global Non-CO_2 Radiative Forcing, 2000-2300 (Post-2100 extrapolations assume constant non-CO_2 radiative forcing after 2100).

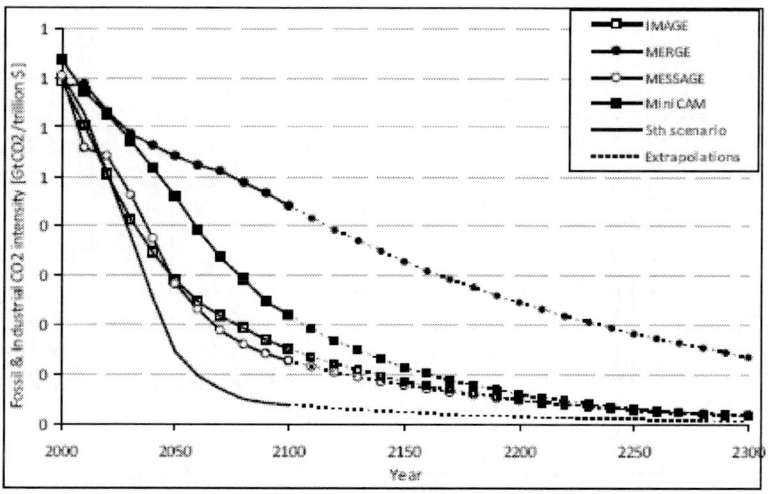

Note: In the fifth scenario, 2000-2100 emissions are equal to the average of the emissions under the 550 ppm CO_2e, full-participation, not-to-exceed scenarios considered by each of the four models.

Figure A7. Global CO_2 Intensity (fossil & industrial CO_2 emissions/GDP), 2000-2300 (Post-2100 extrapolations assume decline in CO_2/GDP growth rate over 2090-2100 is maintained through 2300).

Table A2. 2010 Global SCC Estimates at 2.5 Percent Discount Rate ($2007\$/ton\ CO_2$)

Percentile	1st	5th	10th	25th	50th	Avg	75th	90th	95th	99th
Scenario	PAGE									
IMAGE	3.3	5.9	8.1	13.9	28.8	65.5	68.2	147.9	239.6	563.8
MERGE optimistic	1.9	3.2	4.3	7.2	14.6	34.6	36.2	79.8	124.8	288.3
Message	2.4	4.3	5.8	9.8	20.3	49.2	50.7	114.9	181.7	428.4
MiniCAM base	2.7	4.6	6.4	11.2	22.8	54.7	55.7	120.5	195.3	482.3
5th scenario	2.0	3.5	4.7	8.1	16.3	42.9	41.5	103.9	176.3	371.9
Scenario	DICE									
IMAGE	16.4	21.4	25	33.3	46.8	54.2	69.7	96.3	111.1	130.0
MERGE optimistic	9.7	12.6	14.9	19.7	27.9	31.6	40.7	54.5	63.5	73.3
Message	13.5	17.2	20.1	27	38.5	43.5	55.1	75.8	87.9	103.0
MiniCAM base	13.1	16.7	19.8	26.7	38.6	44.4	56.8	79.5	92.8	109.3
5th scenario	10.8	14	16.7	22.2	32	37.4	47.7	67.8	80.2	96.8
Scenario	FUND									
IMAGE	-33.1	-18.9	-13.3	-5.5	4.1	19.3	18.7	43.5	67.1	150.7
MERGE optimistic	-33.1	-14.8	-10	-3	5.9	14.8	20.4	43.9	65.4	132.9
Message	-32.5	-19.8	-14.6	-7.2	1.5	8.8	13.8	33.7	52.3	119.2
MiniCAM base	-31.0	-15.9	-10.7	-3.4	6	22.2	21	46.4	70.4	152.9
5th scenario	-32.2	-21.6	-16.7	-9.7	-2.3	3	6.7	20.5	34.2	96.8

Table A3. 2010 Global SCC Estimates at 3 Percent Discount Rate ($2007\$/ton\ CO_2$)

Percentile	1st	5th	10th	25th	50th	Avg	75th	90th	95th	99th
Scenario	PAGE									
IMAGE	2.0	3.5	4.8	8.1	16.5	39.5	41.6	90.3	142.4	327.4
MERGE optimistic	1.2	2.1	2.8	4.6	9.3	22.3	22.8	51.3	82.4	190.0
Message	1.6	2.7	3.6	6.2	12.5	30.3	31	71.4	115.6	263.0
MiniCAM base	1.7	2.8	3.8	6.5	13.2	31.8	32.4	72.6	115.4	287.0

Percentile	1st	5th	10th	25th	50th	Avg	75th	90th	95th	99th
Scenario	PAGE									
5th scenario	1.3	2.3	3.1	5	9.6	25.4	23.6	62.1	104.7	222.5
Percentile	1st	5th	10th	25th	50th	Avg	75th	90th	95th	99th
Scenario	DICE									
IMAGE	11.0	14.5	17.2	22.8	31.6	35.8	45.4	61.9	70.8	82.1
MERGE optimistic	7.1	9.2	10.8	14.3	19.9	22	27.9	36.9	42.1	48.8
Message	9.7	12.5	14.7	19	26.6	29.8	37.8	51.1	58.6	67.4
MiniCAM base	8.8	11.5	13.6	18	25.2	28.8	36.9	50.4	57.9	67.8
5th scenario	7.9	10.1	11.8	15.6	21.6	24.9	31.8	43.7	50.8	60.6
Scenario	FUND									
IMAGE	-25.2	-15.3	-11.2	-5.6	0.9	8.2	10.4	25.4	39.7	90.3
MERGE optimistic	-24.0	-12.4	-8.7	-3.6	2.6	8	12.2	27	41.3	85.3
Message	-25.3	-16.2	-12.2	-6.8	-0.5	3.6	7.7	20.1	32.1	72.5
MiniCAM base	-23.1	-12.9	-9.3	-4	2.4	10.2	12.2	27.7	42.6	93.0
5th scenario	-24.1	-16.6	-13.2	-8.3	-3	-0.2	2.9	11.2	19.4	53.6

Table A4. 2010 Global SCC Estimates at 5 Percent Discount Rate ($2007\$/ton\ CO_2$)

Percentile	1st	5th	10th	25th	50th	Avg	75th	90th	95th	99th
Scenario	PAGE									
IMAGE	0.5	0.8	1.1	1.8	3.5	8.3	8.5	19.5	31.4	67.2
MERGE optimistic	0.3	0.5	0.7	1.2	2.3	5.2	5.4	12.3	19.5	42.4
Message	0.4	0.7	0.9	1.6	3	7.2	7.2	17	28.2	60.8
MiniCAM base	0.3	0.6	0.8	1.4	2.7	6.4	6.6	15.9	24.9	52.6
5th scenario	0.3	0.6	0.8	1.3	2.3	5.5	5	12.9	22	48.7
Scenario	DICE									
IMAGE	4.2	5.4	6.2	7.6	10	10.8	13.4	16.8	18.7	21.1
MERGE optimistic	2.9	3.7	4.2	5.3	7	7.5	9.3	11.7	12.9	14.4
Message	3.9	4.9	5.5	7	9.2	9.8	12.2	15.4	17.1	18.8
MiniCAM base	3.4	4.2	4.7	6	7.9	8.6	10.7	13.5	15.1	16.9
5th scenario	3.2	4	4.6	5.7	7.6	8.2	10.2	12.8	14.3	16.0
Scenario	FUND									
IMAGE	-11.7	-8.4	-6.9	-4.6	-2.2	-1.3	0.7	4.1	7.4	17.4
MERGE optimistic	-10.6	-7.1	-5.6	-3.6	-1.3	-0.3	1.6	5.4	9.1	19.0
Message	-12.2	-8.9	-7.3	-4.9	-2.5	-1.9	0.3	3.5	6.5	15.6
MiniCAM base	-10.4	-7.2	-5.8	-3.8	-1.5	-0.6	1.3	4.8	8.2	18.0
5th scenario	-10.9	-8.3	-7	-5	-2.9	-2.7	-0.8	1.4	3.2	9.2

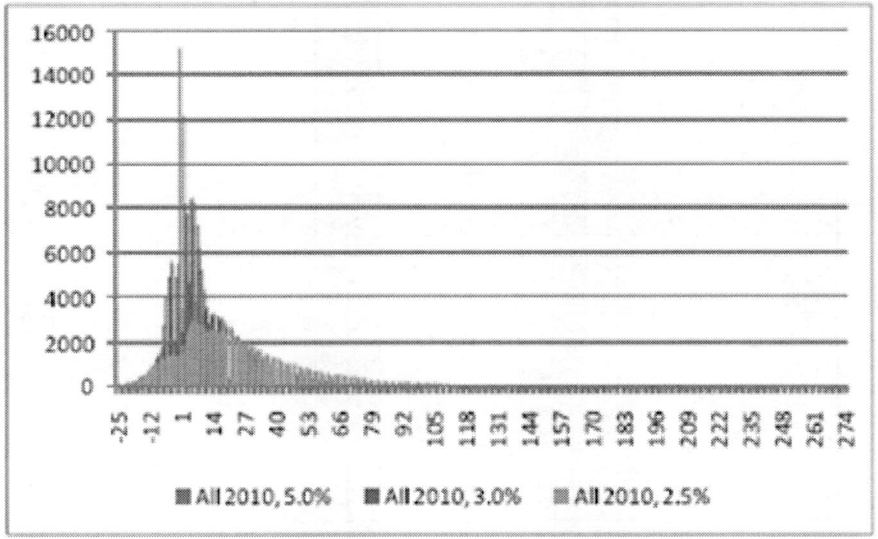

* The distribution of SCC values ranges from -$5,192 to $66,116 but the X-axis has been truncated approximately the 1st and 99th percentiles to better show the data.

Figure A8. Histogram of Global SCC Estimates in 2010 (2007$/ton CO_2), by discount rate.

Table A5. Additional Summary Statistics of 2010 Global SCC Estimates

Discount rate:	5%				3%				2.5%			
Scenario	Mean	Variance	Skewness	Kurtosis	Mean	Variance	Skewness	Kurtosis	Mean	Variance	Skewness	Kurtosis
DICE	9.0	13.1	0.8	0.2	28.3	209.8	1.1	0.9	42.2	534.9	1.2	1.1
PAGE	6.5	136.0	6.3	72.4	29.8	3,383.7	8.6	151.0	49.3	9,546.0	8.7	143.8
FUND	-1.3	70.1	28.2	1,479.0	6.0	16,382.5	128.0	18,976.5	13.6	150,732.6	149.0	23,558.3

REFERENCES

Andronova, N., and M. Schlesinger. 2001. "Objective estimation of the probability density function for climate sensitivity." *J. Geophys. Res.*, 106(D19), 22605–22611.

Annan, J., et al., 2005. "Efficiently constraining climate sensitivity with paleoclimate simulations." *Scientific Online Letters on the Atmosphere*, 1, 181–184.

Anthoff D, C. Hepburn, and R. Tol. 2009a. "Equity Weighting and the Marginal Damage Costs of Climate Change." *Ecological Economics* 68:836-849.

Anthoff, D., R. Tol, and G. Yohe. 2009b. "Risk aversion, time preference, and the social cost of carbon." *Environmental Research Letters* 4: 024002 (7pp).

Arrow, K. 2007. "Global climate change: a challenge to policy." *Economist's Voice* 4(3):Article 2. Arrow, K. 2000. "A Comment on Cooper." *The World Bank Research Observer*. vol 15, no. 2.

Arrow, K., et al. 1996. *Benefit-Cost Analysis in Environmental, Health, and Safety Regulation: A Statement of Principles*. Washington, D.C., AEI Press. pp. 13-14.

Arrow, K.J., et al. 1996. "Intertemporal equity, discounting and economic efficiency," in Climate Change 1995: Economic and Social Dimensions of Climate Change, Contribution of Working Group III to the *Second Assessment Report of the Intergovernmental Panel on Climate Change*.

Campbell, J., P. Diamond, and J. Shoven. 2001. "Estimating the Real Rate of Return on Stocks Over the Long Term." Presented to the Social Security Advisory Board. August.

Campbell, K. et al. 2007. *The age of consequences: The foreign policy and national security implications of global climate change*. Center for Strategic & International Studies, 119 pp.

Castles, I. and D. Henderson. 2003. "The IPCC Emission Scenarios: An Economic-Statistical Critique." *Energy and Environment* 14(2-3): 159-185.

Chetty, R. 2006. "A New Method of Estimating Risk Aversion." *American Economic Review* 96(5): 1821–1834.

Dasgupta P. 2006. "Comments on the Stern Review's economics of climate change." University of Cambridge working paper.

Dasgupta P. 2008. "Discounting climate change." *Journal of Risk and Uncertainty* 37:141-169.

Easterling, W., et al. 2007. *Climate Change 2007: Impacts, Adaptation, and Vulnerability*. Intergovernmental Panel on Climate Change, 976 pp. http://www.ipcc.ch/publications and data/publications ipcc fourth assessment report wg2 report i mpactsadaptationandvulnerability.htm, last accessed March 25, 2010.

Evans D., and H. Sezer. 2005. "Social discount rates for member countries of the European Union." *J. Econ. Stud.* 32 47–59.

Forest, C., et al. 2002. "Quantifying uncertainties in climate system properties with the use of recent observations." *Science* 295, 113.

Forest, D., P. Stone, and A. Sokolov. 2006. "Estimated PDFs of climate system properties including natural and anthropogenic forcings." *Geophys. Res. Lett.*, 33, L01705.

Forster, P., and J. Gregory. 2006. "The climate sensitivity and its components diagnosed from Earth radiation budget data." *J. Clim.*, 19, 39–52.

Frame, D., et al. 2005. "Constraining climate forecasts: The role of prior assumptions." *Geophys. Res. Lett.*, 32, L09702.

Gingerich, P. 2006. "Environment and evolution through the Paleocene-Eocene thermal maximum." *Trends Ecol. Evol.* 21: 246-253.

Gollier, C. 2008. "Discounting with fat-tailed economic growth." *Journal of Risk and Uncertainty* 37:171- 186.

Gollier, C. and M. Weitzman (2009). "How Should the Distant Future be Discounted When Discount Rates are Uncertain?" Harvard University, mimeo, Nov 2009.

Gregory, J., et al. 2002a. "An observationally based estimate of the climate sensitivity." *J. Clim.*, 15(22), 3117–3121.

Groom, B., Koundouri, P., Panipoulou, E., Pantelidis, T. 2006. "An econometric approach to estimating long-run discount rates." *Journal of Applied Econometrics*.

Hall R, and C. Jones . 2007. "The value of life and the rise in health spending." Quarterly *Journal of Economics* 122(1):39-72.

Hansen, J.,M. Sato, P. Kharecha, G. Russell, D. W. Lea and M. Siddall. 2007. "Climate change and trace gases." *Phil. Trans. Roy. Soc. A* 365: 1925-1954.

Hegerl G, et al. 2007. "Understanding and attributing climate change." in Solomon S, et al. (eds) *Climate Change 2007: the physical science basis. Contribution of Working Group I to the Fourth Assessment Report of the Intergovernmental Panel on Climate Change*. Cambridge, UK: Cambridge University Press. http://www.ipcc.ch/publicationsanddata/ar4/wg1/en/contents.html

Hegerl, G., T. Crowley, W. Hyde, and D. Frame. 2006. "Constraints on climate sensitivity from temperature reconstructions of the past seven centuries." *Nature* 440.

Holtsmark, B., and K. Alfsen. 2005. "PPP Correction of the IPCC Emission Scenarios – Does it Matter?" *Climatic Change* 68(1-2): 11-19.

Hope C. 2006. "The marginal impact of CO_2 from PAGE2002: an integrated assessment model incorporating the IPCC's five reasons for concern." *The Integrated Assessment Journal* 6(1):19-56.

Hope C. 2008. "Optimal carbon emissions and the social cost of carbon under uncertainty." *The Integrated Assessment Journal* 8(1):107-122.

Intergovernmental Panel on Climate Change (2007). "Summary for Policymakers." In *Climate Change 2007: The Physical Science Basis, Contribution of Working Group I to the Fourth Assessment Report of the Intergovernmental Panel on Climate Change*. Cambridge University Press.

Just, R., D. Hueth, and A. Schmitz. 2004. *The Welfare Economics of Public Policy*. Glos UK: Edward Elgar Publishing Limited.

Knutti, R., T. Stocker, F. Joos, and G. Plattner. 2002. "Constraints on radiative forcing and future climate change from observations and climate model ensembles." *Nature*, 416, 719–723.

Kriegler, E. et al. 2009. "Imprecise probability assessment of tipping points in the climate system." *Proc. Natl. Acad. Sci.* 106: 5041-5046.

Kotlikoff, L. and D. Rapson. 2006. "Does It Pay, at the Margin, to Work and Save? – Measuring Effective Marginal Taxes on Americans' Labor Supply and Saving." National Bureau of Economic Research, Working Paper, No. 12533.

Le Treut H., et al. 2007. "Historical Overview of Climate Change." in Solomon et al., *Climate Change 2007*.

Lenton, T., et al. 2008. "Tipping elements in the Earth's climate system." *Proc. Natl. Acad. Sci.* 105: 1786- 1793.

Levy, M., et al. 2005. "Ecosystem conditions and human well-being." In: *Ecosystems and Human Wellbeing: Current State and Trends, Volume 1.* [R. Hassan, R. Scholes, and N. Ash, eds.] Washington: Island
Press. pp. 123-164.

Lind, R. 1990. "Reassessing the Government's Discount Rate Policy in Light of New Theory and Data in a World Economy with a High Degree of Capital Mobility." *Journal of Environmental Economics and Management* 18, S-8-S-28.

Mastrandre, M. 2009. "Calculating the benefits of climate policy: Examining the assumptions of Integrated Assessment Models." Pew Center on Global Climate Change Working Paper, 60 pp.

Meehl, G, et al. 2007. "Global Climate Projections." in Solomon et al., *Climate Change 2007*.

National Research Council (2009). *Hidden Costs of Energy: Unpriced Consequences of Energy Production and Use*. National Academies Press.

Newbold S, Daigneault A. 2009. "Climate response uncertainty and the benefits of greenhouse gas emissions reductions." *Environmental and Resource Economics* 44:351-377.

Newell, R., and W. Pizer. 2003. Discounting the distant future: how much do uncertain rates increase valuations? *Journal of Environmental Economics and Management* 46: 52-71.

Nordhaus, W. 1994. "Expert Opinion on Climate Change." *American Scientist* 82: 45-51.

Nordhaus, W. 2007a. *Accompanying notes and documentation on development of DICE-2007 model: notes on DICE-2007.delta.v8 as of September 21, 2007.*

Nordhaus, W. 2007b. "Alternative measures of output in global economic-environmental models: Purchasing power parity or market exchange rates?" *Energy Economics* 29: 349-372.

Nordhaus W. 2008. *A Question of Balance: Weighing the Options on Global Warming Policies*. New Haven, CT: Yale University Press.

Nordhaus, W. 2009. "An Analysis of the Dismal Theorem. Cowles Foundation Discussion Paper. No. 1686. January.

Nordhaus W., and Boyer J. 2000. *Warming the World: Economic Models of Global Warming*. Cambridge, MA: MIT Press.

Pindyck, R. 2009. "Uncertain Outcomes and Climate Change Policy." NBER Working Paper, No. 15259. August.

Ramsey, F. 1928. "A Mathematical Theory of Saving." *The Economic Journal* 38(152): 543-559.

Roe, G. 2008. "Feedbacks, timescales, and seeing red." *Annual Review of Earth and Planetary Sciences* 37:5.1-5.23.

Roe, G., and M. Baker. 2007. "Why is climate sensitivity so unpredictable?" *Science* 318:629-632.

Schneider von Deimling, T., H. Held, A. Ganopolski, and S. Rahmstorf. 2006. "Climate sensitivity estimated from ensemble simulations of glacial climate." *Clim. Dyn.*, 27, 149–163.

Smith, J. et al. 2009. "Transient dwarfism of soil fauna during the Paleocene-Eocene Thermal Maximum." *Proc. Natl. Acad. Sci.* 106: 17665-17660.

Stern, N., et al. (2006), *Stern Review: The Economics of Climate Change*, HM Treasury, London. See http://www.hm-treasury.gov.uk/sternreview report. htm, last accessed March 25, 2010.

Stern N. 2008. "The economics of climate change." *American Economic Review* 98(2):1-37.

Sterner, T., and U. Persson. 2008. An even Sterner review: Introducing relative prices into the discounting debate. *Rev. Env. Econ. Pol.* 2: 61-76.

Summers, L., and R. Zeckhauser. 2008. "Policymaking for Posterity." Journal of Risk and Uncertainty 37: 115-140.

Szpiro, G. 1986. "Measuring Risk Aversion: An Alternative Approach." *The Review of Economics and Statistics* 68(1): 156-9.

Tol, R. 2002a. "Estimates of the damage costs of climate change. Part I: benchmark estimates." *Environmental and Resource Economics* 21:47-73.

Tol, R. 2002b. "Estimates of the damage costs of climate change. Part II: dynamic estimates." *Environmental and Resource Economics* 21:135-160.

Tol, R. 2006. "Exchange Rates and Climate Change: An Application of FUND." *Climatic Change* 75(1-2): 59-80.

Tol, R. 2009. "An analysis of mitigation as a response to climate change." Copenhagen Consensus on Climate. Discussion Paper.

U.S. Department of Defense. 2010. Quadrennial Defense Review Report. February.

Warren, R., et al. 2006. "Spotlight Impact Functions in Integrated Assessment." Tyndall Center for Climate Change Research, Working Paper 91.

Weitzman, M. 2009. "On modeling and interpreting the economics of catastrophic climate change." *Review of Economics and Statistics* 91:1-19.

Weitzman, M. 2007. "A review of The Stern Review of the Economics of Climate Change." *Journal of Economic Literature* 45:703-724.

Weitzman, M. 1999. In Portney P.R. and Weyant J.P. (eds.), *Discounting and Intergenerational Equity*, Resources for the Future, Washington, D.C.

Weitzman, M. 1998. "Why the Far-Distant Future Should Be Discounted at Its Lowest Possible Rate." *Journal of Environmental Economics and Management* 36 (3): 201-208.

Wing, S. et al. 2005. "Transient floral change and rapid global warming at the Paleocene-Eocene boundary." *Science* 310: 993-996.

End Notes

[1] In this document, we present all values of the SCC as the cost per metric ton of CO_2 emissions. Alternatively, one could report the SCC as the cost per metric ton of carbon emissions. The multiplier for translating between mass of CO_2 and the mass of carbon is 3.67 (the molecular weight of CO_2 divided by the molecular weight of carbon = 44/12 = 3.67).

[2] The DICE (Dynamic Integrated Climate and Economy) model by William Nordhaus evolved from a series of energy models and was first presented in 1990 (Nordhaus and Boyer 2000, Nordhaus 2008). The PAGE (Policy Analysis of the Greenhouse Effect) model was developed by Chris Hope in 1991 for use by European decision-makers in assessing the marginal impact of carbon emissions (Hope 2006, Hope 2008). The FUND (Climate Framework for Uncertainty, Negotiation, and Distribution) model, developed by Richard Tol in the early 1990s, originally to study international capital transfers in climate policy. is now widely used to study climate impacts (e.g., Tol 2002a, Tol 2002b, Anthoff et al. 2009, Tol 2009).

[3] Using the default assumptions in DICE 2007, this effect generates an approximately 25 percent increase in the SCC relative to damages calculated by fixing GDP. In DICE2007, the time path of GDP is endogenous. Specifically, the path of GDP depends on the rate of saving and level of abatement in each period chosen by the optimizing representative agent in the model. We made two modifications to DICE to make it consistent with EMF GDP trajectories (see next section): we assumed a fixed rate of savings of 20%, and we re-calibrated the exogenous path of total factor productivity so that DICE would produce GDP projections in the absence of warming that exactly matched the EMF scenarios.

[4] In the deterministic version of FUND, the majority of damages are attributable to increased air conditioning demand, while reduced cold stress in Europe, North America, and Central and East Asia results in health benefits in those regions at low to moderate levels of warming (Warren et al., 2006).

[5] The x-axis represents increases in annual, rather than equilibrium, temperature, while the y-axis represents the annual stream of benefits as a share of global GDP. Each specific combination of climate sensitivity, socioeconomic, and emissions parameters will produce a different realization of damages for each IAM. The damage functions represented in Figures 1A and 1B are the outcome of default assumptions. For instance, under alternate assumptions, the damages from FUND may cross from negative to positive at less than or greater than 3 °C.

[6] It is true that federal statutes are presumed not to have extraterritorial effect, in part to ensure that the laws of the United States respect the interests of foreign sovereigns. But use of a global measure for the SCC does not give extraterritorial effect to federal law and hence does not intrude on such interests.

[7] It is plausible that a loss of $X inflicts more serious harm on a poor nation than on a wealthy one, but development of the appropriate "equity weight" is challenging. Emissions reductions also impose costs, and hence a full account would have to consider that a given cost of emissions reductions imposes a greater utility or welfare loss on a poor nation than on a wealthy one. Even if equity weighting—for both the costs and benefits of emissions reductions—is appropriate when considering the utility or welfare effects of international action, the interagency group concluded that it should not be used in developing an SCC for use in regulatory policy at this time.

[8] Based on 2008 GDP (in current US dollars) from the World Bank Development Indicators Report.

[9] The equilibrium climate sensitivity includes the response of the climate system to increased greenhouse gas concentrations over the short to medium term (up to 100-200 years), but it does not include long-term feedback effects due to possible large-scale changes in ice sheets or the biosphere, which occur on a time scale of many hundreds to thousands of years (e.g. Hansen et al. 2007).

[10] This is in accord with the judgment that it "is likely to lie in the range 2 °C to 4.5 °C" and the IPCC definition of "likely" as greater than 66 percent probability (Le Treut et al.2007). "Very likely" indicates a greater than 90 percent probability.

[11] Strictly speaking, "most likely" refers to the mode of a distribution rather than the median, but common usage would allow the mode, median, or mean to serve as candidates for the central or "most likely" value and the IPCC report is not specific on this point. For the distributions we considered, the median was between the mode and the mean. For the Roe and Baker distribution, setting the median equal to 3°C, rather than the mode or mean, gave a 95th percentile that is more consistent with IPCC judgments and the literature. For example, setting the mean and mode equal to 3°C produced 95th percentiles of 5.6 and 8.6 °C, respectively, which are in the lower and upper end of the range in the literature. Finally, the median is closer to 3°C than is the mode for the truncated distributions selected by the IPCC (Hegerl, et al., 2006); the average median is 3.1 °C and the average mode is 2.3 °C, which is most consistent with a Roe and Baker distribution with the median set equal to 3 °C.

[12] The estimates based on instrumental data are from Andronova and Schlesinger (2001), Forest et al. (2002; dashed line, anthropogenic forcings only), Forest et al. (2006; solid line, anthropogenic and natural forcings), Gregory et al. (2002a), Knutti et al. (2002), Frame et al. (2005), and Forster and Gregory (2006). Hegerl et al. (2006) are based on multiple palaeoclimatic reconstructions of north hemisphere mean temperatures over the last 700 years. Also shown are the 5-95 percent approximate ranges for two estimates from the last glacial maximum (dashed, Annan et al. 2005; solid, Schneider von Deimling et al. 2006), which are based on models with different structural properties.

[13] Such an emissions path would be consistent with widespread action by countries to mitigate GHG emissions, I though it could also result from technological advances. It was chosen because it represents the most stringent case analyzed by the EMF-22 where all the models converge: a 550 ppm, not to exceed, full participation scenario.

[14] While the EMF-22 models used market exchange rates (MER) to calculate global GDP, it is also possible to use purchasing power parity (PPP). PPP takes into account the different price levels across countries, so it more accurately describes relative standards of living across countries. MERs tend to make low-income countries appear poorer than they actually are. Because many models assume convergence in per capita income over time, use of MER-adjusted GDP gives rise to projections of higher economic growth in low income countries. There is an ongoing debate about how much this will affect estimated climate impacts. Critics of the use of MER argue that it leads to overstated economic growth and hence a significant upward bias in projections of greenhouse gas emissions, and unrealistically high future temperatures (e.g., Castles and Henderson 2003). Others argue that convergence of the emissions-intensity gap across countries at least partially offset the overstated income gap so that differences in exchange rates have less of an effect on emissions (Holtsmark and Alfsen, 2005; Tol, 2006). Nordhaus (2007b) argues that the ideal approach is to use superlative PPP accounts (i.e., using cross-sectional PPP measures for relative incomes and outputs and national accounts price and quantity indexes for time-series extrapolations). However, he notes that it important to keep this debate in

perspective; it is by no means clear that exchange-rate-conversion issues are as important as uncertainties about population, technological change, or the many geophysical uncertainties.

[15] For instance, in the MESSAGE model's reference case total primary energy production from nuclear, biomass, and non-biomass renewables is projected to increase from about 15 percent of total primary energy in 2000 to 54 percent in 2100. In comparison, the MiniCAM reference case shows 10 percent in 2000 and 21 percent in 2100.

[16] For example, MiniCAM projects if all non-US OECD countries reduce CO_2 emissions to 83 percent below 2005 levels by 2050 (per the G-8 agreement) but all other countries continue along a BAU path CO_2 concentrations in 2100 would drop from 794 ppmv in its reference case to 762 ppmv.

[17] The literature argues for a risk-free rate on government bonds as an appropriate measure of the consumption rate of interest. Arrow (2000) suggests that it is roughly 3-4 percent. OMB cites evidence of a 3.1 percent pre-tax rate for 10-year Treasury notes in the A-4 guidance. Newell and Pizer (2003) find real interest rates between 3.5 and 4 percent for 30-year Treasury securities.

[18] The positive approach reflects how individuals make allocation choices across time, but it is important to keep in mind that we wish to reflect preferences for society as a whole, which generally has a longer planning horizon.

[19] Cambell et al (2001) estimates that the annual real return from stocks for 1900-1995 was about 7 percent. The annual real rate of return for the S&P 500 from 1950 – 2008 was about 6.8 percent. In the absence of a better way to population-weight the tax rates, we use the middle of the 20 – 40 percent range to derive a post-tax interest rate (Kotlikoff and Rapson 2006).

[20] The parameter p measures the pure rate of time preference: people's behavior reveals a preference for an increase in utility today versus the future. Consequently, it is standard to place a lower weight on utility in the future. The parameter ç captures diminishing marginal utility: consumption in the future is likely to be higher than consumption today, so diminishing marginal utility of consumption implies that the same monetary damage will cause a smaller reduction of utility for wealthier individuals, either in the future or in current generations. If ç= 0, then a one dollar increase in income is equally valuable regardless of level of income; if ç= 1, then a one percent increase in income is equally valuable no matter the level of income; and if ç> 1, then a one percent increase in income is less valuable to wealthier individuals.

[21] In this case, g could be taken from the selected EMF socioeconomic scenarios or alternative assumptions about the rate of consumption growth.

[22] Empirical estimates of ç span a wide range of values. A benchmark value of 2 is near the middle of the range of values estimated or used by Szpiro (1986), Hall and Jones (2007), Arrow (2007), Dasgupta (2006, 2008), Weitzman (2007, 2009), and Nordhaus (2008). However, Chetty (2006) developed a method of estimating ç using data on labor supply behavior. He shows that existing evidence of the effects of wage changes on labor supply imposes a tight upper bound on the curvature of utility over wealth (CRRA < 2) with the mean implied value of 0.71 and concludes that the standard expected utility model cannot generate high levels of risk aversion without contradicting established facts about labor supply. Recent work has jointly estimated the components of the Ramsey equation. Evans and Sezer (2005) estimate q = 1.49 for 22 OECD countries. They also estimate p = 1.08 percent per year using data on mortality rates. Anthoff, et al. (2009b) estimate q = 1.18, and p = 1.4 percent. When they multiply the bivariate probability distributions from their work and Evans and Sezer (2005) together, they find q = 1.47, and p = 1.07.

[23] Stern (2008) argues that building in a positive rate of exogenous technical change over time reduces the implied savings rate and that 11 at or above 2 are inconsistent with observed behavior with regard to equity. (At the same time, adding exogenous technical change—all else equal—would increase g as well.)

[24] For instance, the UK applies a discount rate of 3.5 percent to the first 30 years; 3 percent for years 31 - 75; 2.5 percent for years 76 - 125; 2 percent for years 126 - 200; 1.5 percent for years 201 - 300; and 1 percent after 300 years. As a sensitivity, it recommends a discount rate of 3 percent for the first 30 years, also decreasing over time.

[25] Uncertainty in future damages is distinct from uncertainty in the discount rate. Weitzman (2008) argues that Stern's choice of a low discount rate was "right for the wrong reasons." He demonstrates how the damages from a low probability, catastrophic event far in the future dominate the effect of the discount rate in a present value calculation and result in an infinite willingness-to-pay for mitigation today. Newbold and Daigneault, (2009) and Nordhaus (2009) find that Weitzman's result is sensitive to the functional forms chosen for climate sensitivity, utility, and consumption. Summers and Zeckhauser (2008) argue that uncertainty in future damages can also work in the other direction by increasing the benefits of waiting to learn the appropriate level of mitigation required.

[26] Calculations done by Pizer et al. using the original simulation program from Newell and Pizer (2003).

[27] Nordhaus (2008) runs DICE2007 with $\rho = 1.5$ and $q = 2$. The default approach in PAGE2002 (version 1.4epm) treats ρ and q as random parameters, specified using a triangular distribution such that the min, mode, and max = 0.1, 1, and 2 for ρ, and 0.5, 1, and 2 for q, respectively. The FUND default value for q is 1, and Tol generates SCC estimates for values of $\rho = 0$, 1, and 3 in many recent papers (e.g. Anthoff et al. 2009). The path of per-capita consumption growth, g, varies over time but is treated deterministically in two of the three models. In DICE, g is endogenous. Under Ramsey discounting, as economic growth slows in the future, the large damages from climate change that occur far out in the future are discounted at a lower rate than impacts that occur in the nearer term.

[28] However, it is possible that other benefits or costs of proposed regulations unrelated to CO_2 emissions will be discounted at rates that differ from those used to develop the SCC estimates.

[29] However these research dollars will be diverted from whatever their next best use would have been in the absence of climate change (so productivity/GDP would have been still higher).

[30] Note EMF did not provide CO_2 concentrations for the IMAGE reference scenario. Thus, for this scenario, we fed the fossil, industrial and land CO_2 emissions into MAGICC (considered a "neutral arbiter" model, which is tuned to emulate the major global climate models) and the resulting CO_2 concentrations were used. Note also that MERGE assumes a neutral biosphere so net land CO_2 emissions are set to zero for all years for the MERGE Optimistic reference scenario, and for the MERGE component of the average 550 scenario (i.e., we add up the land use emissions from the other three models and divide by 4).

[31] Both the model default CH4 emissions and the initial atmospheric CH4 is set to zero to avoid double counting the effect of past CH4 emissions.

[32] AR4 Synthesis Report, p. 44, http://www.ipcc.ch/pdf/assessment-report/ar4/syr/ar4 syr.pdf

[33] See Smith, S.J., R. Andres, E. Conception, and J. Lurz, 2004: Historical sulfur dioxide emissions, 1850-2000: LI methods and results. Joint Global Research Institute, College Park, 14 pp.

[34] See Carmichael, G., D. Streets, G. Calori, M. Amann, M. Jacobson, J. Hansen, and H. Ueda, 2002: Changing trends LI in sulphur emissions in Asia: implications for acid deposition, air

pollution, and climate. Environmental Science and LI Technology, 36(22):4707- 4713; Streets, D., K. Jiang, X. Hu, J. Sinton, X.-Q. Zhang, D. Xu, M. Jacobson, and J. LI Hansen, 2001: Recent reductions in China's greenhouse gas emissions. Science, 294(5548): 1835-1837.

[35] United Nations. 2004. World Population to 2300. http://www.un.org/esa/population/publications/longrange2/worldpop2300final.pdf

[36] MERGE assumes a neutral biosphere so net land CO_2 emissions are set to zero for all years for the MERGE Optimistic reference scenario, and for the MERGE component of the average 550 scenario (i.e., we add up the land use emissions from the other three models and divide by 4).

In: Social Cost of Carbon Estimates ... ISBN: 978-1-63463-691-9
Editor: Joseph M. Keck

Chapter 2

TECHNICAL SUPPORT DOCUMENT: TECHNICAL UPDATE OF THE SOCIAL COST OF CARBON FOR REGULATORY IMPACT ANALYSIS UNDER EXECUTIVE ORDER 12866*

Interagency Working Group on Social Cost of Carbon

EXECUTIVE SUMMARY

Under Executive Order 12866, agencies are required, to the extent permitted by law, "to assess both the costs and the benefits of the intended regulation and, recognizing that some costs and benefits are difficult to quantify, propose or adopt a regulation only upon a reasoned determination that the benefits of the intended regulation justify its costs." The purpose of the "social cost of carbon" (SCC) estimates presented here is to allow agencies to incorporate the social benefits of reducing carbon dioxide (CO_2) emissions into cost-benefit analyses of regulatory actions that impact cumulative global emissions. The SCC is an estimate of the monetized damages associated with an incremental increase in carbon emissions in a given year. It is intended to

* This is an edited, reformatted and augmented version of a report issued May 2013.

include (but is not limited to) changes in net agricultural productivity, human health, property damages from increased flood risk, and the value of ecosystem services due to climate change.

The interagency process that developed the original U.S. government's SCC estimates is described in the 2010 interagency technical support document (TSD) (Interagency Working Group on Social Cost of Carbon 2010). Through that process the interagency group selected four SCC values for use in regulatory analyses. Three values are based on the average SCC from three integrated assessment models (IAMs), at discount rates of 2.5, 3, and 5 percent. The fourth value, which represents the 95th percentile SCC estimate across all three models at a 3 percent discount rate, is included to represent higher-than-expected impacts from temperature change further out in the tails of the SCC distribution.

While acknowledging the continued limitations of the approach taken by the interagency group in 2010, this document provides an update of the SCC estimates based on new versions of each IAM (DICE, PAGE, and FUND). It does not revisit other interagency modeling decisions (e.g., with regard to the discount rate, reference case socioeconomic and emission scenarios, or equilibrium climate sensitivity). Improvements in the way damages are modeled are confined to those that have been incorporated into the latest versions of the models by the developers themselves in the peer-reviewed literature.

The SCC estimates using the updated versions of the models are higher than those reported in the 2010 TSD. By way of comparison, the four 2020 SCC estimates reported in the 2010 TSD were \$7, \$26, \$42 and \$81 (2007\$). The corresponding four updated SCC estimates for 2020 are \$12, \$43, \$65, and \$129 (2007\$). The model updates that are relevant to the SCC estimates include: an explicit representation of sea level rise damages in the DICE and PAGE models; updated adaptation assumptions, revisions to ensure damages are constrained by GDP, updated regional scaling of damages, and a revised treatment of potentially abrupt shifts in climate damages in the PAGE model; an updated carbon cycle in the DICE model; and updated damage functions for sea level rise impacts, the agricultural sector, and reduced space heating requirements, as well as changes to the transient response of temperature to the buildup of GHG concentrations and the inclusion of indirect effects of methane emissions in the FUND model. The SCC estimates vary by year, and the following table summarizes the revised SCC estimates from 2010 through 2050.

Revised Social Cost of CO_2, 2010 – 2050 (in 2007 dollars per metric ton of CO_2)

Discount Rate Year	5.0% Avg	3.0% Avg	2.5% Avg	3.0% 95th
2010	11	33	52	90
2015	12	38	58	109
2020	12	43	65	129
2025	14	48	70	144
2030	16	52	76	159
2035	19	57	81	176
2040	21	62	87	192
2045	24	66	92	206
2050	27	71	98	221

I. PURPOSE

The purpose of this document is to update the schedule of social cost of carbon (SCC) estimates from the 2010 interagency technical support document (TSD) (Interagency Working Group on Social Cost of Carbon 2010).[1] E.O. 13563 commits the Administration to regulatory decision making "based on the best available science."[2] Additionally, the interagency group recommended in 2010 that the SCC estimates be revisited on a regular basis or as model updates that reflect the growing body of scientific and economic knowledge become available.[3] New versions of the three integrated assessment models used by the U.S. government to estimate the SCC (DICE, FUND, and PAGE), are now available and have been published in the peer reviewed literature. While acknowledging the continued limitations of the approach taken by the interagency group in 2010 (documented in the original 2010 TSD), this document provides an update of the SCC estimates based on the latest peer-reviewed version of the models, replacing model versions that were developed up to ten years ago in a rapidly evolving field. It does not revisit other assumptions with regard to the discount rate, reference case socioeconomic and emission scenarios, or equilibrium climate sensitivity. Improvements in the way damages are modeled are confined to those that have been incorporated into the latest versions of the models by the developers themselves in the peer-reviewed literature. The agencies participating in the interagency working group continue to investigate potential improvements to the way in which economic damages associated with changes in CO_2 emissions are quantified.

Section II summarizes the major updates relevant to SCC estimation that are contained in the new versions of the integrated assessment models released since the 2010 interagency report. Section III presents the updated schedule of SCC estimates for 2010 – 2050 based on these versions of the models. Section IV provides a discussion of other model limitations and research gaps.

II. SUMMARY OF MODEL UPDATES

This section briefly summarizes changes to the most recent versions of the three integrated assessment models (IAMs) used by the interagency group in 2010. We focus on describing those model updates that are relevant to estimating the social cost of carbon, as summarized in Table 1. For example, both the DICE and PAGE models now include an explicit representation of sea level rise damages. Other revisions to PAGE include: updated adaptation assumptions, revisions to ensure damages are constrained by GDP, updated regional scaling of damages, and a revised treatment of potentially abrupt shifts in climate damages. The DICE model's simple carbon cycle has been updated to be more consistent with a more complex climate model. The FUND model includes updated damage functions for sea level rise impacts, the agricultural sector, and reduced space heating requirements, as well as changes to the transient response of temperature to the buildup of GHG concentrations and the inclusion of indirect effects of methane emissions. Changes made to parts of the models that are superseded by the interagency working group's modeling assumptions – regarding equilibrium climate sensitivity, discounting, and socioeconomic variables – are not discussed here but can be found in the references provided in each section below.

A. Dice

DICE 2010 includes a number of changes over the previous 2007 version used in the 2010 interagency report. The model changes that are relevant for the SCC estimates developed by the interagency working group include: 1) updated parameter values for the carbon cycle model, 2) an explicit representation of sea level dynamics, and 3) a re-calibrated damage function that includes an explicit representation of economic damages from sea level rise. Changes were also made to other parts of the DICE model—including the equilibrium climate sensitivity parameter, the rate of change of total factor productivity, and the elasticity of the marginal utility of consumption—but these components of DICE are superseded by the interagency working group's assumptions and so will not be discussed here. More details on DICE2007 can be found in Nordhaus (2008) and on DICE2010 in Nordhaus (2010). The DICE2010 model and documentation is also available for download from the homepage of William Nordhaus.

Table 1. Summary of Key Model Revisions Relevant to the Interagency SCC

IAM	Version used in 2010 Interagency Analysis	New Version	Key changes relevant to interagency SCC
DICE	2007	2010	Updated calibration of the carbon cycle model and explicit representation of sea level rise (SLR) and associated damages.
FUND	3.5	3.8	Updated damage functions for space heating, SLR,
	(2009)	(2012)	agricultural impacts, changes to transient response of temperature to buildup of GHG concentrations, and inclusion of indirect climate effects of methane.
PAGE	2002	2009	Explicit representation of SLR damages, revisions to damage function to ensure damages do not exceed100% of GDP, change in regional scaling of damages, revised treatment of potential abrupt damages, and updated adaptation assumptions.

Carbon Cycle Parameters

DICE uses a three-box model of carbon stocks and flows to represent the accumulation and transfer of carbon among the atmosphere, the shallow ocean and terrestrial biosphere, and the deep ocean. These parameters are "calibrated to match the carbon cycle in the Model for the Assessment of Greenhouse Gas Induced Climate Change (MAGICC)" (Nordhaus 2008 p 44).[4] Carbon cycle transfer coefficient values in DICE2010 are based on re-calibration of the model to match the newer 2009 version of MAGICC (Nordhaus 2010 p 2). For example, in DICE2010, in each decade, 12 percent of the carbon in the atmosphere is transferred to the shallow ocean, 4.7 percent of the carbon in the shallow ocean is transferred to the atmosphere, 94.8 percent remains in the shallow ocean, and 0.5 percent is transferred to the deep ocean. For comparison, in DICE 2007, 18.9 percent of the carbon in the atmosphere is transferred to the shallow ocean each decade, 9.7 percent of the carbon in the shallow ocean is transferred to the atmosphere, 85.3 percent remains in the shallow ocean, and 5 percent is transferred to the deep ocean.

The implication of these changes for DICE2010 is in general a weakening of the ocean as a carbon sink and therefore a higher concentration of carbon in the atmosphere than in DICE2007, for a given path of emissions. All else equal, these changes will generally increase the level of warming and therefore the SCC estimates in DICE2010 relative to those from DICE2007.

Sea Level Dynamics

A new feature of DICE2010 is an explicit representation of the dynamics of the global average sea level anomaly to be used in the updated damage function (discussed below). This section contains a brief description of the sea level rise (SLR) module; a more detailed description can be found on the model developer's website.[5] The average global sea level anomaly is modeled as the sum of four terms that represent contributions from: 1) thermal expansion of the oceans, 2) melting of glaciers and small ice caps, 3) melting of the Greenland ice sheet, and 4) melting of the Antarctic ice sheet.

The parameters of the four components of the SLR module are calibrated to match consensus results from the IPCC's Fourth Assessment Report (AR4).[6] The rise in sea level from thermal expansion in each time period (decade) is 2 percent of the difference between the sea level in the previous period and the long run equilibrium sea level, which is 0.5 meters per degree Celsius (°C) above the average global temperature in 1900. The rise in sea level from the melting of glaciers and small ice caps occurs at a rate of 0.008 meters per decade per °C above the average global temperature in 1900.

The contribution to sea level rise from melting of the Greenland ice sheet is more complex. The equilibrium contribution to SLR is 0 meters for temperature anomalies less than 1 °C and increases linearly from 0 meters to a maximum of 7.3 meters for temperature anomalies between 1 °C and 3.5 °C. The contribution to SLR in each period is proportional to the difference between the previous period's sea level anomaly and the equilibrium sea level anomaly, where the constant of proportionality increases with the temperature anomaly in the current period.

The contribution to SLR from the melting of the Antarctic ice sheet is -0.001 meters per decade when the temperature anomaly is below 3 °C and increases linearly between 3 °C and 6 °C to a maximum rate of 0.025 meters per decade at a temperature anomaly of 6 °C.

Re-calibrated Damage Function

Economic damages from climate change in the DICE model are represented by a fractional loss of gross economic output in each period. A portion of the remaining economic output in each period (net of climate change damages) is consumed and the remainder is invested in the physical capital stock to support future economic production, so each period's climate damages will reduce consumption in that period and in all future periods due to the lost investment. The fraction of output in each period that is lost due to climate change impacts is represented as one minus a fraction, which is one divided by a quadratic function of the temperature anomaly, producing a sigmoid ("S"-shaped) function.[7] The loss function in DICE2010 has been expanded by adding a quadratic function of SLR to the quadratic function of temperature. In DICE2010 the temperature anomaly coefficients have been recalibrated to avoid double-counting damages from sea level rise that were implicitly included in these parameters in DICE2007.

The aggregate damages in DICE2010 are illustrated by Nordhaus (2010 p 3), who notes that "...damages in the uncontrolled (baseline) [i.e., reference] case ... in 2095 are $12 trillion, or 2.8 percent of global output, for a global temperature increase of 3.4 °C above 1900 levels." This compares to a loss of 3.2 percent of global output at 3.4 °C in DICE2007. However, in DICE2010, annual damages are lower in most of the early periods of the modeling horizon but higher in later periods than would be calculated using the DICE2007 damage function. Specifically, the percent difference between damages in the base run of DICE2010 and those that would be calculated using the DICE2007 damage function starts at +7 percent in 2005, decreases to a low of -14 percent in 2065, then continuously increases to +20 percent by 2300 (the end of the interagency analysis time horizon), and to +160 percent by the end of the model time horizon in 2595. The large increases in the far future years of the time horizon are due to the permanence associated with damages from sea level rise, along with the assumption that the sea level is projected to continue to rise long after the global average temperature begins to decrease. The changes to the loss function generally decrease the interagency working group SCC estimates slightly given that relative increases in damages in later periods are discounted more heavily, all else equal.

B. Fund

FUND version 3.8 includes a number of changes over the previous version 3.5 (Narita et al. 2010) used in the 2010 interagency report. Documentation supporting FUND and the model's source code for all versions of the model is available from the model authors.[8] Notable changes, due to their impact on the SCC estimates, are adjustments to the space heating, agriculture, and sea level rise damage functions in addition to changes to the temperature response function and the inclusion of indirect effects from methane emissions.[9] We discuss each of these in turn.

Space Heating

In FUND, the damages associated with the change in energy needs for space heating are based on the estimated impact due to one degree of warming. These baseline damages are scaled based on the forecasted temperature anomaly's deviation from the one degree benchmark and adjusted for changes in vulnerability due to economic and energy efficiency growth. In FUND 3.5, the function that scales the base year damages adjusted for vulnerability allows for the possibility that in some simulations the benefits associated with reduced heating needs may be an unbounded convex function of the temperature anomaly. In FUND 3.8, the form of the scaling has been modified to ensure that the function is everywhere concave and that there will exist an upper bound on the benefits a region may receive from reduced space heating needs. The new formulation approaches a value of two in the limit of large temperature anomalies, or in other words, assuming no decrease in vulnerability, the reduced expenditures on space heating at any level of warming will not exceed two times the reductions experienced at one degree of warming. Since the reduced need for space heating represents a benefit of climate change in the model, or a negative damage, this change will increase the estimated SCC. This update accounts for a significant portion of the difference in the expected SCC estimates reported by the two versions of the model when run probabilistically.

Sea Level Rise and Land Loss

The FUND model explicitly includes damages associated with the inundation of dry land due to sea level rise. The amount of land lost within a region is dependent upon the proportion of the coastline being protected by adequate sea walls and the amount of sea level rise. In FUND 3.5 the function defining the potential land lost in a given year due to sea level rise is linear in the rate of sea level rise for that year. This assumption implicitly assumes that all regions are well represented by a homogeneous coastline in length and a constant uniform slope moving inland. In FUND 3.8 the function defining the potential land lost has been changed to be a convex function of sea level rise, thereby assuming that the slope of the shore line increases moving inland. The effect of this change is to typically reduce the vulnerability of some regions to sea level rise based land loss, thereby lowering the expected SCC estimate. [10]

Agriculture

In FUND, the damages associated with the agricultural sector are measured as proportional to the sector's value. The fraction is bounded from above by one and is made up of three additive components that represent the effects from carbon fertilization, the rate of temperature change, and the level of the temperature anomaly. In both FUND 3.5 and FUND 3.8, the fraction of the sector's value lost due to the level of the temperature anomaly is modeled as a quadratic function with an intercept of zero. In FUND 3.5, the coefficients of this loss function are modeled as the ratio of two random normal variables. This specification had the potential for unintended extreme behavior as draws from the parameter in the denominator approached zero or went negative. In FUND 3.8, the coefficients are drawn directly from truncated normal distributions so that they remain in the range and, respectively,ensuring the correct sign and eliminating the potential for divide by zero errors. The means for the new distributions are set equal to the ratio of the means from the normal distributions used in the previous version. In general the impact of this change has been to decrease the range of the distribution while spreading out the distributions' mass over the remaining range relative to the previous version. The net effect of this change on the SCC estimates is difficult to predict.

Transient Temperature Response

The temperature response model translates changes in global levels of radiative forcing into the current expected temperature anomaly. In FUND, a given year's increase in the temperature anomaly is based on a mean reverting function where the mean equals the equilibrium temperature anomaly that would eventually be reached if that year's level of radiative forcing were sustained. The rate of mean reversion defines the rate at which the transient temperature approaches the equilibrium. In FUND 3.5, the rate of temperature response is defined as a decreasing linear function of equilibrium climate sensitivity to capture the fact that the progressive heat uptake of the deep ocean causes the rate to slow at higher values of the equilibrium climate sensitivity. In FUND 3.8, the rate of temperature response has been updated to a quadratic function of the equilibrium climate sensitivity. This change reduces the sensitivity of the rate of temperature response to the level of the equilibrium climate sensitivity, a relationship first noted by Hansen et al. (1985) based on the heat uptake of the deep ocean. Therefore in FUND 3.8, the temperature response will typically be faster than in the previous version. The overall effect of this change is likely to increase estimates of the SCC as higher temperatures are reached during the timeframe analyzed and as the same damages experienced in the previous version of the model are now experienced earlier and therefore discounted less.

Methane

The IPCC AR4 notes a series of indirect effects of methane emissions, and has developed methods for proxying such effects when computing the global warming potential of methane (Forster et al. 2007). FUND 3.8 now includes the same methods for incorporating the indirect effects of methane emissions. Specifically, the average atmospheric lifetime of methane has been set to 12 years to account for the feedback of methane emissions on its own lifetime. The radiative forcing associated with atmospheric methane has also been increased by 40% to account for its net impact on ozone production and stratospheric water vapor. All else equal, the effect of this increased radiative forcing will be to increase the estimated SCC values, due to greater projected temperature anomaly.

C. Page

PAGE09 (Hope 2013) includes a number of changes from PAGE2002, the version used in the 2010 SCC interagency report. The changes that most directly affect the SCC estimates include: explicitly modeling the impacts from sea level rise, revisions to the damage function to ensure damages are constrained by GDP, a change in the regional scaling of damages, a revised treatment for the probability of a discontinuity within the damage function, and revised assumptions on adaptation. The model also includes revisions to the carbon cycle feedback and the calculation of regional temperatures.[11] More details on PAGE09 can be found in Hope (2011a, 2011b, 2011c). A description of PAGE2002 can be found in Hope (2006).

Sea Level Rise

While PAGE2002 aggregates all damages into two categories – economic and non-economic impacts -, PAGE09 adds a third explicit category: damages from sea level rise. In the previous version of the model, damages from sea level rise were subsumed by the other damage categories. In PAGE09 sea level damages increase less than linearly with sea level under the assumption that land, people, and GDP are more concentrated in low-lying shoreline areas. Damages from the economic and non-economic sector were adjusted to account for the introduction of this new category.

Revised Damage Function to Account for Saturation

In PAGE09, small initial economic and non-economic benefits (negative damages) are modeled for small temperature increases, but all regions eventually experience economic damages from climate change, where damages are the sum of additively separable polynomial functions of temperature and sea level rise. Damages transition from this polynomial function to a logistic path once they exceed a certain proportion of remaining Gross Domestic Product (GDP) to ensure that damages do not exceed 100 percent of GDP. This differs from PAGE2002, which allowed Eastern Europe to potentially experience large benefits from temperature increases, and which also did not bound the possible damages that could be experienced.

Regional Scaling Factors

As in the previous version of PAGE, the PAGE09 model calculates the damages for the European Union (EU) and then, assumes that damages for other regions are proportional based on a given scaling factor. The scaling factor in PAGE09 is based on the length of a region's coastline relative to the EU (Hope 2011b). Because of the long coastline in the EU, other regions are, on average, less vulnerable than the EU for the same sea level and temperature increase, but all regions have a positive scaling factor. PAGE2002 based its scaling factors on four studies reported in the IPCC's third assessment report, and allowed for benefits from temperature increase in Eastern Europe, smaller impacts in developed countries, and higher damages in developing countries.

Probability of a Discontinuity

In PAGE2002, the damages associated with a "discontinuity" (nonlinear extreme event) were modeled as an expected value. Specifically, a stochastic probability of a discontinuity was multiplied by the damages associated with a discontinuity to obtain an expected value, and this was added to the economic and non-economic impacts. That is, additional damages from an extreme event, such as extreme melting of the Greenland ice sheet, were multiplied by the probability of the event occurring and added to the damage estimate. In PAGE09, the probability of discontinuity is treated as a discrete event for each year in the model. The damages for each model run are estimated either with or without a discontinuity occurring, rather than as an expected value. A large-scale discontinuity becomes possible when the temperature rises beyond some threshold value between 2 and 4°C. The probability that a discontinuity will occur beyond this threshold then increases by between 10 and 30 percent for every 1°C rise in temperature beyond the threshold. If a discontinuity occurs, the EU loses an additional 5 to 25 percent of its GDP (drawn from a triangular distribution with a mean of 15 percent) in addition to other damages, and other regions lose an amount determined by the regional scaling factor. The threshold value for a possible discontinuity is lower than in PAGE2002, while the rate at which the probability of a discontinuity increases with the temperature anomaly and the damages that result from a discontinuity are both higher than in PAGE2002. The model assumes that only one discontinuity can occur and that the impact is phased in over a period of time, but once it occurs, its effect is permanent.

Adaptation

As in PAGE2002, adaptation is available to help mitigate any climate change impacts that occur. In PAGE this adaptation is the same regardless of the temperature change or sea level rise and is therefore akin to what is more commonly considered a reduction in vulnerability. It is modeled by reducing the damages by some percentage. PAGE09 assumes a smaller decrease in vulnerability than the previous version of the model and assumes that it will take longer for this change in vulnerability to be realized. In the aggregated economic sector, at the time of full implementation, this adaptation will mitigate all damages up to a temperature increase of 1°C, and for temperature anomalies between 1°C and 2°C, it will reduce damages by 15-30 percent (depending on the region). However, it takes 20 years to fully implement this adaptation. In PAGE2002, adaptation was assumed to reduce economic sector damages up to 2°C by 50-90 percent after 20 years. Beyond 2°C, no adaptation is assumed to be available to mitigate the impacts of climate change. For the non-economic sector, in PAGE09 adaptation is available to reduce 15 percent of the damages due to a temperature increase between 0°C and 2°C and is assumed to take 40 years to fully implement, instead of 25 percent of the damages over 20 years assumed in PAGE2002. Similarly, adaptation is assumed to alleviate 25-50 percent of the damages from the first 0.20 to 0.25 meters of sea level rise but is assumed to be ineffective thereafter. Hope (2011c) estimates that the less optimistic assumptions regarding the ability to offset impacts of temperature and sea level rise via adaptation increase the SCC by approximately 30 percent.

Other Noteworthy Changes

Two other changes in the model are worth noting. There is a change in the way the model accounts for decreased CO_2 absorption on land and in the ocean as temperature rises. PAGE09 introduces a linear feedback from global mean temperature to the percentage gain in the excess concentration of CO_2, capped at a maximum level. In PAGE2002, an additional amount was added to the CO_2 emissions each period to account for a decrease in ocean absorption and a loss of soil carbon. Also updated is the method by which the average global and annual temperature anomaly is downscaled to determine annual average regional temperature anomalies to be used in the regional damage functions. In PAGE2002, the scaling was determined solely based on regional difference in emissions of sulfate aerosols. In PAGE09, this regional temperature anomaly is further adjusted using an additive factor that is based on the average absolute latitude of a region relative to the area weighted average absolute latitude of the Earth's landmass, to capture relatively greater changes in temperature forecast to be experienced at higher latitudes.

III. Revised SCC Estimates

The updated versions of the three integrated assessment models were run using the same methodology detailed in the 2010 TSD (Interagency Working Group on Social Cost of Carbon 2010). The approach along with the inputs for the socioeconomic emissions scenarios, equilibrium climate sensitivity distribution, and discount rate remains the same. This includes the five reference scenarios based on the EMF-22 modeling exercise, the Roe and Baker equilibrium climate sensitivity distribution calibrated to the IPCC AR4, and three constant discount rates of 2.5, 3, and 5 percent.

As was previously the case, the use of three models, three discount rates, and five scenarios produces 45 separate distributions for the global SCC. The approach laid out in the 2010 TSD applied equal weight to each model and socioeconomic scenario in order to reduce the dimensionality down to three separate distributions representative of the three discount rates. The interagency group selected four values from these distributions for use in regulatory analysis. Three values are based on the average SCC across models and socio-economic-emissions scenarios at the 2.5, 3, and 5 percent discount rates, respectively. The fourth value was chosen to represent the higher-than-expected economic impacts from climate change further out in the tails of the SCC distribution. For this purpose, the 95th percentile of the SCC estimates at a 3 percent discount rate was chosen. (A detailed set of percentiles by model and scenario combination and additional summary statistics for the 2020 values is available in the Appendix.) As noted in the 2010 TSD, "the 3 percent discount rate is the central value, and so the central value that emerges is the average SCC across models at the 3 percent discount rate" (Interagency Working Group on Social Cost of Carbon 2010, p. 25). However, for purposes of capturing the uncertainties involved in regulatory impact analysis, the interagency group emphasizes the importance and value of including all four SCC values.

Table 2 shows the four selected SCC estimates in five year increments from 2010 to 2050. Values for 2010, 2020, 2030, 2040, and 2050 are calculated by first combining all outputs (10,000 estimates per model run) from all scenarios and models for a given discount rate. Values for the years in between are calculated using linear interpolation. The full set of revised annual SCC estimates between 2010 and 2050 is reported in the Appendix.

Table 2. Revised Social Cost of CO_2, 2010 – 2050 (in 2007 dollars per metric ton of CO_2)

Discount Rate Year	5.0% Avg	3.0% Avg	2.5% Avg	3.0% 95th
2010	11	33	52	90
2015	12	38	58	109
2020	12	43	65	129
2025	14	48	70	144
2030	16	52	76	159
2035	19	57	81	176
2040	21	62	87	192
2045	24	66	92	206
2050	27	71	98	221

The SCC estimates using the updated versions of the models are higher than those reported in the 2010 TSD due to the changes to the models outlined in the previous section. By way of comparison, the 2020 SCC estimates reported in the original TSD were $7, $26, $42 and $81 (2007$) (Interagency

Working Group on Social Cost of Carbon 2010). Figure 1 illustrates where the four SCC values for 2020 fall within the full distribution for each discount rate based on the combined set of runs for each model and scenario (150,000 estimates in total for each discount rate). In general, the distributions are skewed to the right and have long tails. The Figure also shows that the lower the discount rate, the longer the right tail of the distribution.

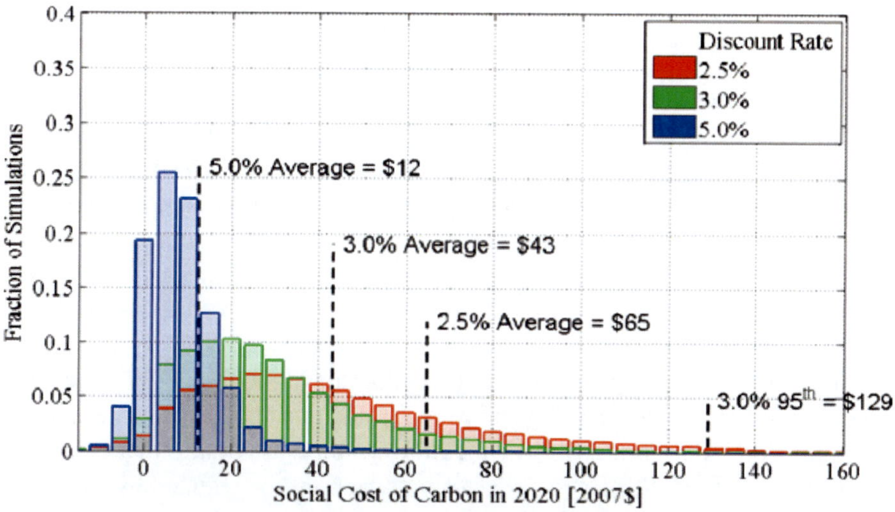

Figure 1. Distribution of SCC Estimates for 2020 (in 2007$ per metric ton CO_2).

As was the case in the 2010 TSD, the SCC increases over time because future emissions are expected to produce larger incremental damages as physical and economic systems become more stressed in response to greater climatic change. The approach taken by the interagency group is to compute the cost of a marginal ton emitted in the future by running the models for a set of perturbation years out to 2050. Table 3 illustrates how the growth rate for these four SCC estimates varies over time.

The future monetized value of emission reductions in each year (the SCC in year *t* multiplied by the change in emissions in year *t*) must be discounted to the present to determine its total net present value for use in regulatory analysis. As previously discussed in the 2010 TSD, damages from future emissions should be discounted at the same rate as that used to calculate the SCC estimates themselves to ensure internal consistency – i.e., future damages from climate change, whether they result from emissions today or emissions in a later year, should be discounted using the same rate.

Table 3. Average Annual Growth Rates of SCC Estimates between 2010 and 2050

Average Annual Growth Rate (%)	5.0% Avg	3.0% Avg	2.5% Avg	3.0% 95th
2010-2020	1.2%	3.2%	2.4%	4.3%
2020-2030	3.4%	2.1%	1.7%	2.4%
2030-2040	3.0%	1.8%	1.5%	2.0%
2040-2050	2.6%	1.6%	1.3%	1.5%

Under current OMB guidance contained in Circular A-4, analysis of economically significant proposed and final regulations from the domestic perspective is required, while analysis from the international perspective is optional. However, the climate change problem is highly unusual in at least two respects. First, it involves a global externality: emissions of most greenhouse gases contribute to damages around the world even when they are emitted in the United States. Consequently, to address the global nature of the problem, the SCC must incorporate the full (global) damages caused by GHG emissions. Second, climate change presents a problem that the United States alone cannot solve. Even if the United States were to reduce its greenhouse gas emissions to zero, that step would be far from enough to avoid substantial climate change. Other countries would also need to take action to reduce emissions if significant changes in the global climate are to be avoided. Emphasizing the need for a global solution to a global problem, the United States has been actively involved in seeking international agreements to reduce emissions and in encouraging other nations, including emerging major economies, to take significant steps to reduce emissions. When these considerations are taken as a whole, the interagency group concluded that a global measure of the benefits from reducing U.S. emissions is preferable. For additional discussion, see the 2010 TSD.

IV. Other Model Limitations and Research Gaps

The 2010 interagency SCC TSD discusses a number of important limitations for which additional research is needed. In particular, the document highlights the need to improve the quantification of both non-catastrophic and catastrophic damages, the treatment of adaptation and technological change, and the way in which inter-regional and inter-sectoral linkages are modeled.

While the new version of the models discussed above offer some improvements in these areas, further work remains warranted. The 2010 TSD also discusses the need to more carefully assess the implications of risk aversion for SCC estimation as well as the inability to perfectly substitute between climate and non-climate goods at higher temperature increases, both of which have implications for the discount rate used. EPA, DOE, and other agencies continue to engage in research on modeling and valuation of climate impacts that can potentially improve SCC estimation in the future.

APPENDIX

Table A1. Annual SCC Values: 2010-2050 (2007$/metric ton CO_2)

Discount Rate	5.0%	3.0%	2.5%	3.0%
Year	Avg	Avg	Avg	95th
2010	11	33	52	90
2011	11	34	54	94
2012	11	35	55	98
2013	11	36	56	102
2014	11	37	57	106
2015	2	38	58	109
2016	12	39	60	113
2017	12	40	61	117
2018	12	41	62	121
2019	12	42	63	125
2020	12	43	65	129
2021	13	44	66	132
2022	13	45	67	135
2023	13	46	68	138
2024	14	47	69	141
2025	14	48	70	144
2026	15	49	71	147
2027	15	49	72	150
2028	15	50	73	153
2029	16	51	74	156
2030	16	52	76	159
2031	17	53	77	163
2032	17	54	78	166
2033	18	55	79	169
2034	18	56	80	172

Table A1. (Continued)

Discount Rate	5.0%	3.0%	2.5%	3.0%
2035	19	57	81	176
2036	19	58	82	179
2037	20	59	84	182
2038	20	60	85	185
2039	21	61	86	188
2040	21	62	87	192
2041	22	63	88	195
2042	22	64	89	198
2043	23	65	90	200
2044	23	65	91	203
2045	24	66	92	206
2046	24	67	94	209
2047	25	68	95	212
2048	25	69	96	215
2049	26	70	97	218
2050	27	71	98	221

Table A2. 2020 Global SCC Estimates at 2.5 Percent Discount Rate ($2007\$/metric\ ton\ CO_2$)

Percentile	1st	5th	10th	25th	50th	Avg	75th	90th	95th	99th
Scenario12						PAGE				
IMAGE	6	11	15	27	58	129	139	327	515	991
MERGE Optimistic	4	6	9	16	34	78	82	196	317	649
MESSAGE	4	8	11	20	42	108	107	278	483	918
MiniCAM Base	5	9	12	22	47	107	113	266	431	872
5th Scenario	2	4	6	11	25	85	68	200	387	955
Scenario						DICE				
IMAGE	25	31	37	47	64	72	92	123	139	161
MERGE Optimistic	14	18	20	26	36	40	50	65	74	85
MESSAGE	20	24	28	37	51	58	71	95	109	221
MiniCAM Base	20	25	29	38	53	61	76	102	117	135
5th Scenario	17	22	25	33	45	52	65	91	106	126
Scenario						FUND				
IMAGE	-17	-1	5	17	34	44	59	90	113	176
MERGE Optimistic	-7	2	7	16	30	35	49	72	91	146
MESSAGE	-19	-4	2	12	27	32	46	70	87	135
MiniCAM Base	-9	1	8	18	35	45	59	87	108	172
5th Scenario	-30	-12	-5	6	19	24	35	57	72	108

Table A3. 2020 Global SCC Estimates at 3 Percent Discount Rate (2007$/metric ton CO_2)

Percentile	1st	5th	10th	25th	50th	Avg	75th	90th	95th	99th
Scenario					PAGE					
IMAGE	4	7	10	18	38	91	95	238	385	727
MERGE Optimistic	2	4	6	11	23	56	58	142	232	481
MESSAGE	3	5	7	13	29	75	74	197	330	641
MiniCAM Base	3	5	8	14	30	73	75	184	300	623
5th Scenario	1	3	4	7	17	58	48	136	264	660
Scenario					DICE					
IMAGE	16	21	24	32	43	48	60	79	90	102
MERGE Optimistic	10	13	15	19	25	28	35	44	50	58
MESSAGE	14	18	20	26	35	40	49	64	73	83
MiniCAM Base	13	17	20	26	35	39	49	65	73	85
5th Scenario	12	15	17	22	30	34	43	58	67	79
Scenario					FUND					
IMAGE	-14	-3	1	9	20	25	35	54	69	111
MERGE Optimistic	-8	-1	3	9	18	22	31	47	60	97
MESSAGE	-16	-5	-1	6	16	18	28	43	55	88
MiniCAM Base	-9	-1	3	10	21	27	35	53	67	107
5th Scenario	-22	-10	-5	2	10	13	20	33	42	63

Table A4. 2020 Global SCC Estimates at 5 Percent Discount Rate (2007$/metric ton CO_2)

Percentile	1st	5th	10th	25th	50th	Avg	75th	90th	95th	99th
Scenario					PAGE					
IMAGE	1	2	2	5	10	28	27	71	123	244
MERGE Optimistic	1	1	2	3	7	17	17	45	75	153
MESSAGE	1	1	2	4	9	24	22	60	106	216
MiniCAM Base	1	1	2	3	8	21	21	54	94	190
5th Scenario	0	1	1	2	5	18	14	41	78	208
Scenario					DICE					
IMAGE	6	8	9	11	14	15	18	22	25	27
MERGE Optimistic	4	5	6	7	9	10	12	15	16	18
MESSAGE	6	7	8	10	12	13	16	20	22	25
MiniCAMBase	5	6	7	8	11	12	14	18	20	22
5th Scenario	5	6	6	8	10	11	14	17	19	21

Table A4. (Continued)

Percentile	1st	5th	10th	25th	50th	Avg	75th	90th	95th	99th
Scenario					FUND					
IMAGE	-9	-5	-3	-1	2	3	6	11	15	25
MERGE Optimistic	-6	-3	-2	0	3	4	7	12	16	27
MESSAGE	-10	-6	-4	-1	2	2	5	9	13	23
MiniCAM Base	-7	-3	-2	0	3	4	7	11	15	26
5th Scenario	-11	-7	-5	-2	0	0	3	6	8	14

Table A5. Additional Summary Statistics of 2020 Global SCC Estimates

Discount rate:	5.0%				3.0%				2.5%			
Statistic:	Mean	Variance	Skewness	Kurtosis	Mean	Variance	Skewness	Kurtosis	Mean	Variance	Skewness	Kurtosis
DICE	12	26	2	15	38	409	3	24	57	1097	3	30
PAGE	22	1616	5	32	71	14953	4	22	101	29312	4	23
FUND	3	560	-170	35222	21	22487	-85	18842	36	68055	-46	13105

References

Anthoff, D. and Tol, R.S.J. 2013. The uncertainty about the social cost of carbon: a decomposition analysis using FUND. *Climatic Change* 117: 515–530.

Fankhauser, S. 1995. Valuing climate change: The economics of the greenhouse. London, England: Earthscan.

Forster, P., V. Ramaswamy, P. Artaxo, T. Berntsen, R. Betts, D.W. Fahey, J. Haywood, J. Lean, D.C. Lowe, G. Myhre, J. Nganga, R. Prinn, G. Raga, M. Schulz and R. Van Dorland. 2007. Changes in Atmospheric Constituents and in Radiative Forcing. In: *Climate Change 2007: The Physical Science Basis. Contribution of Working Group I to the Fourth Assessment Report of the Intergovernmental Panel on Climate Change* [Solomon, S., D. Qin, M. Manning, Z. Chen, M. Marquis, K.B. Averyt, M.Tignor and H.L. Miller (eds.)]. Cambridge University Press, Cambridge, United Kingdom and New York, NY, USA.

Hope, Chris. 2006. "The Marginal Impact of CO2 from PAGE2002: An Integrated Assessment Model Incorporating the IPCC's Five Reasons for Concern." *The Integrated Assessment Journal*. 6(1): 19–56.

Hope, Chris. 2011a "The PAGE09 Integrated Assessment Model: A Technical Description" Cambridge Judge Business School Working Paper No. 4/2011 (April). Accessed November 23, 2011: http://www.jbs.cam.ac.uk/research/workingpapers/2011/wp1104.pdf.

Hope, Chris. 2011b "The Social Cost of CO2 from the PAGE09 Model" Cambridge Judge Business School

Working Paper No. 5/2011 (June). Accessed November 23, 2011: http://www.jbs.cam.ac.uk/research/workingpapers/2011/wp1105.pdf.

Hope, Chris. 2011c "New Insights from the PAGE09 Model: The Social Cost of CO2" Cambridge Judge Business School Working Paper No. 8/2011 (July). Accessed November 23, 2011: http://www.jbs.cam.ac.uk/research/working papers/2011/wp1108.pdf.

Hope, C. 2013. Critical issues for the calculation of the social cost of CO2: why the estimates from PAGE09 are higher than those from PAGE2002. *Climatic Change* 117: 531–543.

Interagency Working Group on Social Cost of Carbon. 2010. Social Cost of Carbon for Regulatory Impact Analysis under Executive Order 12866. February. United States Government. http://www.whitehouse.gov/sites/default/files/omb/inforeg/for-agencies/Social-Cost-of-Carbon-forRIA.pdf.

Meehl, G.A., T.F. Stocker, W.D. Collins, P. Friedlingstein, A.T. Gaye, J.M. Gregory, A. Kitoh, R. Knutti, J.M. Murphy, A. Noda, S.C.B. Raper, I.G. Watterson, A.J. Weaver and Z.-C. Zhao. 2007. Global Climate Projections. In: *Climate Change 2007: The Physical Science Basis. Contribution of Working Group I to the Fourth Assessment Report of the Intergovernmental Panel on Climate Change* [Solomon, S., D. Qin, M. Manning, Z. Chen, M. Marquis, K.B. Averyt, M. Tignor and H.L. Miller (eds.)]. Cambridge University Press, Cambridge, United Kingdom and New York, NY, USA.

Narita, D., R. S. J. Tol and D. Anthoff. 2010. Economic costs of extratropical storms under climate change: an application of FUND. *Journal of Environmental Planning and Management* 53(3): 371-384.

National Academy of Sciences. 2011. *Climate Stabilization Targets: Emissions, Concentrations, and Impacts over Decades to Millennia.* Washington, DC: National Academies Press, Inc.

Nicholls, R.J., N. Marinova, J.A. Lowe, S. Brown, P. Vellinga, D. de Gusmão, J. Hinkel and R.S.J. Tol. 2011. Sea-level rise and its possible impacts given a 'beyond 4°C world' in the twenty-first century. *Phil. Trans. R. Soc. A* 369(1934): 161-181.

Nordhaus, W. 2010. Economic aspects of global warming in a post-Copenhagen environment. *Proceedings of the National Academy of Sciences* 107(26): 11721-11726.

Nordhaus, W. 2008. A Question of Balance: Weighing the Options on Global Warming Policies. New Haven, CT: Yale University Press.

Randall, D.A., R.A. Wood, S. Bony, R. Colman, T. Fichefet, J. Fyfe, V. Kattsov, A. Pitman, J. Shukla, J. Srinivasan, R.J. Stouffer, A. Sumi and K.E. Taylor. 2007. Climate Models and Their Evaluation. In: *Climate Change 2007: The Physical Science Basis. Contribution of Working Group I to the Fourth Assessment Report of the Intergovernmental Panel on Climate Change* [Solomon, S., D. Qin, M. Manning, Z. Chen, M. Marquis, K.B. Averyt, M.Tignor and H.L. Miller (eds.)]. Cambridge University Press, Cambridge, United Kingdom and New York, NY, USA.

End Notes

[1] In this document, we present all values of the SCC as the cost per metric ton of CO2 emissions. Alternatively, one could report the SCC as the cost per metric ton of carbon emissions. The multiplier for translating between mass of CO2 and the mass of carbon is 3.67 (the molecular weight of CO2 divided by the molecular weight of carbon = 44/12 = 3.67).

[2] http://www.whitehouse.gov/sites/default/files/omb/inforeg/eo12866/eo 135 63_01182011.pdf

[3] See p. 1, 3, 4, 29, and 33 (Interagency Working Group on Social Cost of Carbon 2010).

[4] MAGICC is a simple climate model initially developed by the U.S. National Center for Atmospheric Research that has been used heavily by the Intergovernmental Panel on Climate Change (IPCC) to emulate projections from more sophisticated state of the art earth system simulation models (Randall et al. 2007).

[5] Documentation on the new sea level rise module of DICE is available on William Nordhaus' website at: http://nordhaus.econ.yale.edu/documents/SLR_021910.pdf.

[6] For a review of post-IPCC AR4 research on sea level rise, see Nicholls et al. (2011) and NAS (2011).

[7] The model and documentation, including formulas, are available on the author's webpage at http://www.econ.yale.edu/~nordhaus/homepage/RICEmodels.htm.

[8] http://www.fund-model.org/. This report uses version 3.8 of the FUND model, which represents a modest update to the most recent version of the model to appear in the literature (version 3.7) (Anthoff and Tol, 2013). For the purpose of computing the SCC, the relevant changes (between 3.7 to 3.8) are associated with improving consistency with IPCC AR4 by adjusting the atmospheric lifetimes of CH4 and N2O and incorporating the indirect forcing effects of CH4, along with making minor stability improvements in the sea wall construction algorithm.

[9] The other damage sectors (water resources, space cooling, land loss, migration, ecosystems, human health, and extreme weather) were not significantly updated.

[10] For stability purposes this report also uses an update to the model which assumes that regional coastal protection measures will be built to protect the most valuable land first, such that the

marginal benefits of coastal protection is decreasing in the level of protection following Fankhauser (1995).

[11] Because several changes in the PAGE model are structural (e.g., the addition of sea level rise and treatment of discontinuity), it is not possible to assess the direct impact of each change on the SCC in isolation as done for the other two models above.

In: Social Cost of Carbon Estimates … ISBN: 978-1-63463-691-9
Editor: Joseph M. Keck

Chapter 3

REGULATORY IMPACT ANALYSIS: DEVELOPMENT OF SOCIAL COST OF CARBON ESTIMATES*

United States Government Accountability Office

ABBREVIATIONS

ADS-B	Automatic Dependent Surveillance-Broadcast
ATC	Air Traffic Control
DICE	Dynamic Integrated Climate and Economy
DOT	Department of Transportation
EOP	Executive Office of the President
EPA	Environmental Protection Agency
FUND	Climate Framework for Uncertainty, Negotiation, and Distribution
NHTSA	National Highway Traffic Safety Administration
OMB	Office of Management and Budget
PAGE	Policy Analysis of the Greenhouse Effect

* This is an edited, reformatted and augmented version of a report, GAO-14-663, issued by the Government Accountability Office, Report to Congressional Requesters, dated July 2014.

WHY GAO DID THIS STUDY

Executive Order 12866 directs federal agencies to assess the economic effects of their proposed significant regulatory actions, including a determination that a regulation's benefits justify the costs. In 2008, a federal appeals court directed DOT to update a regulatory impact analysis with an estimate of the social cost of carbon—the dollar value of the net effects (damages and benefits) of an increase in emissions of carbon dioxide, a greenhouse gas.

In 2009, the Interagency Working Group on Social Cost of Carbon was convened to develop estimates for use governmentwide, and it issued final estimates in its 2010 Technical Support Document. In 2013, the group issued revised estimates that were about 50 percent higher than the 2010 estimates, which raised public interest.

GAO was asked to review the working group's development of social cost of carbon estimates. This report describes the participating entities and processes and methods they used to develop the 2010 and 2013 estimates. GAO reviewed executive orders, OMB guidance, the Technical Support Document, its 2013 update, and other key documents. GAO interviewed officials who participated in the working group on behalf of the EOP offices and agencies involved. GAO did not evaluate the quality of the working group's approach.

GAO is making no recommendations in this report. Of seven agencies, OMB and Treasury provided written or oral comments and generally agreed with the findings in this report. Other agencies provided technical comments only or had no comments.

WHAT GAO FOUND

To develop the 2010 and 2013 social cost of carbon estimates, the Office of Management and Budget (OMB) and Council of Economic Advisers convened and led an informal interagency working group in which four other offices from the Executive Office of the President (EOP) and six federal agencies participated. Participating agencies were the Environmental Protection Agency (EPA) and the Departments of Agriculture, Commerce, Energy, Transportation (DOT), and the Treasury. According to several working group participants, the working group included relevant subject-matter experts and the agencies likely to use the estimates in future rulemakings. According to OMB staff, there is no single approach for convening informal interagency working groups and no requirement that this type of working group should document its activities or proceedings. However, OMB and EPA participants stated that the working group documented all major issues discussed in the Technical Support Document, which is consistent with federal standards for internal control. According to the Technical Support Document and participants GAO interviewed, the working group's processes and methods reflected the following three principles:

- *Used consensus-based decision making.* The working group used a consensus-based approach for making key decisions in developing the 2010 and 2013 estimates. Participants generally stated that they were satisfied that the Technical Support Document addressed individual comments on draft versions and reflected the overall consensus of the working group.
- *Relied on existing academic literature and models.* The working group relied largely on existing academic literature and models to develop its estimates. Specifically, the working group used three prevalent academic models that integrate climate and economic data to estimate future economic effects from climate change. The group agreed on three modeling inputs reflecting the wide uncertainty in the academic literature, including discount rates. Once the group reached agreement, EPA officials—sometimes with the assistance of the model developers—calculated the estimates. All other model assumptions and features were unchanged by the working group, which weighted each model equally to calculate estimates. After the academic models were updated to reflect new scientific information, such as in sea level rise and associated damages, the working group used the updated models to revise its estimates in 2013, resulting in higher estimates.
- *Took steps to disclose limitations and incorporate new information.* The Technical Support Document discloses several limitations of the estimates and areas that the working group identified as being in need of additional research. It also sets a goal of revisiting the estimates when substantially updated models become available. Since 2008, agencies have published dozens of regulatory actions for public comment that use various social cost of carbon estimates in regulatory

analyses and, according to working group participants, agencies received many comments on the estimates throughout this process. Several participants told GAO that the working group decided to revise the estimates in 2013 after a number of public comments encouraged revisions because the models used to develop the 2010 estimates had been updated and used in peer-reviewed academic literature.

* * *

July 24, 2014

The Honorable David Vitter
Ranking Member
Committee on Environment and Public Works
United States Senate

The Honorable Tim Murphy
Chairman
Subcommittee on Oversight and Investigations
Committee on Energy and Commerce
House of Representatives

The Honorable John Culberson
House of Representatives

The Honorable Duncan Hunter
House of Representatives

To encourage a regulatory system that protects and improves health, safety, the environment, and the economy, without imposing unreasonable costs on society, federal agencies are required to assess the economic effects of proposed significant regulatory actions. Agencies can use regulatory impact analysis to assess whether a proposed regulation's benefits justify the costs. For example, regulations aimed at benefiting society by decreasing health risks associated with air pollution may require regulated entities, such as power plants, to incur costs for installing pollution control technologies. According to Environmental Protection Agency (EPA) officials, beginning in 2008, some agencies' regulatory impact analyses incorporated estimates of the social cost of carbon,[1] which agencies use to value the net effects of reducing or increasing carbon dioxide emissions.[2] In 2009, in part because agencies used varying estimates of the social cost of carbon, the Executive Office of the President's (EOP) Office of Management and Budget (OMB) and Council of Economic Advisers convened an interagency working group to develop social cost of carbon estimates for federal agencies to use in their regulatory impact analyses. The working group finalized its estimates in 2010 and included them in a document—called the Technical Support Document—that also provides guidance for agencies on using the estimates.[3] In May 2013, the working group issued an update to the Technical Support Document that included revised estimates of the social cost of carbon.[4] These 2013 estimates of the social cost of carbon were approximately 50 percent higher than the 2010 estimates, which raised public interest.

You asked us to review the interagency working group's development of social cost of carbon estimates. This report describes the approach used, including participating entities and processes and methods, to develop the 2010 and 2013 social cost of carbon estimates for regulatory impact analysis.

To address this objective, we reviewed pertinent requirements and guidance, including executive orders and OMB guidance; the Technical Support Document and its 2013 update; published materials and presentations by working group participants on the development of the social cost of carbon estimates; and related GAO reports. We interviewed current and former federal officials or staff who participated in the working group on behalf of the EOP offices and agencies named in the Technical Support Document.[5] We identified these participants by contacting all of the agencies and OMB and then following up with additional individuals identified during our discussions with them. Through this process, we interviewed over 20 individuals who participated in the working group to develop the estimates in the Technical Support Document or its 2013 update, or both. We also corresponded with researchers who developed key academic materials the working group used. Our review describes the approach the working group used to develop estimates of the social cost of carbon; evaluating the quality of the approach is outside the scope of this review.

We conducted this performance audit from November 2013 to July 2014 in accordance with generally accepted government auditing standards. Those standards require that we plan and perform the audit to obtain sufficient, appropriate evidence to provide a reasonable basis for our findings and conclusions based on our audit objectives. We believe that the evidence obtained provides a reasonable basis for our findings and conclusions based on our audit objectives.

BACKGROUND

Executive Order 12866 directs federal agencies to assess the potential costs and benefits of their significant regulatory actions, consisting of several categories of regulatory actions, including those likely to result in a rule that may have an annual effect on the economy of $100 million or more or that have a material adverse effect on the economy; a sector of the economy; productivity; competition; jobs; the environment; public health or safety; or state, local, or tribal governments or communities.[6] Under the executive order, for regulatory actions expected to meet this threshold, or economically significant regulatory actions, agencies must also assess costs and benefits of reasonably feasible alternatives and explain why the planned regulatory action is preferable to the identified alternatives. For each significant regulatory action, the agency is to develop the proposed regulation and associated regulatory impact analysis and submit them to OMB for formal review. After OMB concludes its review, the agency is to publish the proposed rule in the *Federal Register* for public comment. The agency is to issue a document summarizing its consideration of the public comments and, if appropriate, modify the proposed rule in response to the comments. This phase of regulatory development may also include further internal and external review. For significant regulatory actions, the agency is to submit the final regulatory impact analysis and regulation to OMB for review before it publishes the final rule.

In 2003, OMB issued Circular A-4 to provide guidance to federal agencies on the development of regulatory analysis as directed by Executive Order 12866.[7] Circular A-4 states that it is designed to assist agencies by defining good regulatory analysis and standardizing the way benefits and costs of federal regulatory actions are measured and reported. In particular, the guidance provides for systematic evaluation of qualitative and quantitative benefits and costs, including their monetization. Circular A-4 also provides guidance on the selection of discount rates to adjust the estimated benefits and costs for differences in timing.[8] According to Circular A-4, a regulatory impact analysis should include an evaluation of the benefits and costs of the proposed action and any reasonable alternatives, as well as a description of assumptions and uncertainty.[9] It acknowledges that agencies cannot analyze all regulations according to a formula, and that different regulations may call for different emphases in the analysis. Executive Order 13563, which reaffirmed and supplemented Executive Order 12866 in 2011, generally directs federal agencies to conduct regulatory actions based on the best available science.[10] It also directs agencies to use the best available techniques to quantify benefits and costs accurately.

Federal agencies began including estimates of the social cost of carbon in regulatory impact analyses following a decision by the U.S. Court of Appeals for the Ninth Circuit. Specifically, in 2006, the Department of Transportation's National Highway Traffic Safety Administration (NHTSA) issued a final rule on fuel economy standards for certain vehicles which, like other regulations at the time, did not include estimates of the social cost of carbon.[11] The final rule stated that the agency had identified a benefit from a significant reduction in carbon dioxide emissions but stated that the dollar value of the benefit could not be determined because of the wide variation in published estimates of the social cost of carbon. In 2008, in response to a challenge from 11 states and several other organizations, the Ninth Circuit held that NHTSA had acted arbitrarily and capriciously by failing to monetize the value of carbon emissions reduction and directed NHTSA to include such a monetized value in an updated regulatory impact analysis for the regulation.[12] The court stated that, "[w]hile the record shows that there is a range of values, the value of carbon emissions reduction is certainly not zero."[13] Following the court's decision, the Department of Energy, the Department of Transportation, and EPA incorporated a variety of individually developed estimates of the social cost of carbon into their regulatory analyses. These estimates were derived from academic literature and ranged, in general, from $0 to $159 (in 2006, 2007, or 2008 dollars) per metric ton of carbon dioxide emitted in 2007. They also varied in whether they reflected domestic or global measures of the social cost of carbon.[14]

In early 2009, in part to improve consistency in agencies' use of social cost of carbon estimates for regulatory impact analysis, OMB's Office of Information and Regulatory Affairs and the Council of Economic Advisers convened the Interagency Working Group on Social Cost of Carbon. The working group developed interim governmentwide social cost of carbon estimates based on an average of selected estimates published in academic literature. The interim estimates first appeared—and, thus, were first available for public review—in August 2009 in the Department of Energy's final rule on energy standards for vending machines.[15] Agencies subsequently incorporated the interim estimates into several published regulatory actions that sought public comments to inform the development of final estimates for future use. The middle or "central" value for the range of interim estimates was $19 (in 2006 dollars) per metric ton of carbon dioxide emitted in 2007.[16]

Table 1. Central Values for the Social Cost of Carbon Estimates Issued by the Interagency Working Group on Social Cost of Carbon in 2010 and 2013

Dollars are 2007 dollars per metric ton of carbon dioxide

Year	2010 central values	2013 central values
2010	$21	$32
2020	26	43
2030	33	52
2040	39	61
2050	$45	$71

Source: Interagency Working Group on Social Cost of Carbon's Technical Support Document and 2013 update. | GAO-14-663

Note: The Technical Support Document states that the working group calculated the social cost of carbon for emissions occurring in multiple future years to cover the time horizons anticipated for upcoming regulatory analysis. When the benefits and costs of a regulatory action will occur in the future, agencies must determine the present value of future benefits and costs by applying an appropriate discount rate—the interest rate used to convert benefits and costs occurring in different periods to a common present value. According to the Technical Support Document, the social cost of carbon estimates increase over time because future emissions are expected to produce larger incremental damages as the environment and the economy become more stressed in response to greater climate change. The working group selected four values of the social cost of carbon for regulatory analysis. The first three values are based on the average of estimates calculated at discount rates of 2.5 percent, 3 percent, and 5 percent, and the fourth value was included to represent higher-than-expected economic impacts at the 3 percent discount rate. The Technical Support Document refers to the average of estimates calculated at the 3 percent discount rate as the "central value" of the social cost of carbon and states that agencies should consider all four values when conducting regulatory analyses.

In October 2009, after developing the interim estimates, the working group reassembled to begin developing the final social cost of carbon estimates issued in the Technical Support Document. While the Technical Support Document is dated February 2010, it was first released publicly in March 2010 as an appendix to the Department of Energy's final rule on energy standards for small electric motors.[17] Subsequently, dozens of published regulatory actions incorporated the estimates. The Technical Support Document states that the working group agreed to regularly update the social cost of carbon estimates as the research underlying the estimates evolves. In June 2013, after using the 2010 estimates in an earlier proposal of the rule, the Department of Energy's final rule on energy standards for microwaves was the first regulatory action to incorporate the revised estimates developed by the working group in the 2013 update to the Technical Support Document.[18] Table 1 shows the central values for the range of 2010 and 2013 social cost of carbon estimates for carbon emissions occurring in selected years.

Appendix I lists regulatory actions from 2008 to 2014 and the type of social cost of carbon estimates (i.e., individually developed, interim, 2010, or 2013) incorporated in the actions' regulatory analyses.

Approach Used to Develop Estimates of the Social Cost of Carbon

According to the Technical Support Document and participants we interviewed, the working group consisted of participants representing six EOP offices and six federal agencies and was convened under Executive Order 12866. The working group's processes and methods for developing the estimates reflected three key principles. Specifically, according to participants, the working group (1) used consensus-based decision making; (2) relied largely on existing academic literature and models, including technical assistance from outside resources; and (3) took steps to disclose limitations and incorporate new information by considering public comments and revising the estimates as updated research became available.

Participating Entities

According to the Technical Support Document and participants we spoke with, OMB and the Council of Economic Advisers convened and led the working group, and four other EOP offices and six federal agencies actively participated in the group. According to several participants, the participating EOP offices included the relevant subject-matter experts to best contribute on behalf of the EOP,[19] and the other participating agencies were those likely to conduct rulemakings affecting carbon emissions and, therefore, use the social cost of carbon estimates in the future. For example, EPA and the Department of Energy have issued numerous rules using the social cost of carbon estimates (see app. I).

OMB staff and EPA officials told us that OMB and the Council of Economic Advisers decided which EOP offices and federal agencies to invite to participate in the working group and, according to participants we interviewed from several agencies, each agency that chose to participate decided which of its internal offices would send representatives. OMB staff stated that any federal agency was welcome to participate in the working group, and EPA officials told us that at least two invited agencies declined to participate. OMB staff recalled that the working group generally included up to several participants from each participating office and agency and numbered approximately two dozen participants in total.

Table 2 lists the 12 participating offices and agencies, along with the internal offices they sent to represent them on the working group.

Table 2. Offices and Agencies Participating in the Interagency Working Group on Social Cost of Carbon to Develop the 2010 and 2013 Social Cost of Carbon Estimates

	Participating office or agency	*2010 estimates*	*2013 estimates*
Executive Office of the President			
	Council of Economic Advisers[a]	X	X
	Council on Environmental Quality	X	X
	National Economic Council	X	X
	Office of Energy and Climate Change[b]	X	X
	Office of Management and Budget[a] Office of Information and Regulatory Affairs	X	X
	Office of Science and Technology Policy	X	X
Federal agencies			
	Department of Agriculture		
	•Office of the Chief Economist	X	X
	Department of Commerce[c]		
	•International Trade Administration, Office of Competition and Economic Analysis[d]	X	
	•National Oceanic and Atmospheric Administration, National Marine Fisheries Service[e]		X

	Participating office or agency	*2010 estimates*	*2013 estimates*
	Department of Energy[f]		
	•Office of Climate Change Policy and Technology[g]	X	X
	Department of Transportation		
	•Office of the Secretary	X	X
	•Volpe, The National Transportation Systems Center	X	
	Department of the Treasury		
	•Office of Economic Policy	X	
	•Office of International Affairs, Office of Environment and Energy	X	X
	Environmental Protection Agency		
	•Office of Air and Radiation, Office of Atmospheric Programs	X	X
	•Office of Policy, National Center for Environmental Economics	X	X

Source: GAO analysis of information provided by the Office of Management and Budget, Environmental Protection Agency, and Departments of Agriculture, Commerce, Energy, Transportation, and the Treasury. | GAO-14-663.

[a] The Council of Economic Advisers and the Office of Management and Budget convened and led the working group to develop the 2010 and 2013 estimates.

[b] In March 2011, the Office of Energy and Climate Change joined the Domestic Policy Council.

[c] An official from the Department of Commerce's Economics and Statistics Administration told us that he attended two working group meetings as an observer during the development of the 2013 estimates, but that he did not review any materials produced by the group or otherwise contribute to the development of the estimates.

[D] The International Trade Administration's Office of Competition and Economic Analysis is now known as the Office of Trade and Economic Analysis.

[E] An official from the Department of Commerce's National Oceanic and Atmospheric Administration told us that she participated in the working group to develop the 2010 estimates while serving on detail to the Office of Science and Technology Policy.

[f] A former Administrator of the Energy Information Administration told us that he participated as a technical advisor to the working group to develop the 2010 estimates and not as a representative of the Department of Energy. Participants told us that the Energy Information Administration also sent a representative to some working group meetings as an observer during the development of the 2013 estimates.

[g] The Office of Climate Change Policy and Technology is now known as the Office of Climate, Environment, and Efficiency.

In establishing the working group, several participants told us that OMB and the Council of Economic Advisers made efforts to ensure that the group's members, collectively, brought the necessary technical expertise for developing social cost of carbon estimates. For example, according to these participants and EPA documentation, participants from the EOP offices included individuals with expertise in pertinent topics, such as economics and climate science. The former Deputy Assistant Secretary for Environment and Energy at the Department of the Treasury stated that he was invited to participate in the working group because of his prior experience researching ways to discount costs and benefits across generations. In addition, the former Administrator of the Energy Information Administration told us that he was asked to participate, in part, based on his previous experience evaluating climate models while conducting research with the National Academy of Sciences. According to an OMB staff member, the six participating federal agencies were also responsible for ensuring that they provided adequate technical expertise to the working group. Agency representatives included environmental economists and climate scientists, among other key professionals. According to EPA documentation, participants from EPA also provided technical expertise in climate science, economics, and academic modeling to the broader group, as needed.

When the working group reconvened in 2013 to update the estimates, the same EOP offices and agencies generally participated, although some of the individuals participating on behalf of offices or agencies changed, in part due to individuals changing positions or leaving the government altogether. Also, some participants who previously had been serving details at other participating agencies had returned to their home agencies. For example, certain participants who were on detail to the Council of Economic Advisers during the development of the 2010 Technical Support Document instead represented EPA on the working group during the development of the 2013 update.

According to the Technical Support Document, the working group was convened under the broad direction of Executive Order 12866 for agencies to assess the costs and benefits of intended regulations.[20] In addition, participants from several agencies told us that the executive order was the key requirement driving the working group's effort to develop social cost of carbon estimates. OMB staff stated that, while there is no single requirement or other approach for convening interagency working groups, it is appropriate for OMB to form interagency working groups to collaborate on policy or analytic needs identified under Executive Order 12866. These OMB staff members said that, instead of being organized under a written agreement or other requirements, the working group was an informal interagency working group with no charter or other convening document. According to OMB staff, there was no requirement that the informal working group should document its activities or proceedings, including the meetings held or specific discussions that occurred at each. However, OMB staff and EPA officials stated that all major issues discussed during working group meetings are documented in the Technical Support Document and its 2013 update, which is consistent with the control activities standard in the federal standards for internal control.[21] We have also reported that interagency working groups use a variety of mechanisms to implement interagency collaborative efforts, including temporary working groups,[22] and that not all collaborative arrangements, particularly those that are informal, need to be documented through written guidance and agreements.[23]

Processes and Methods

Participants told us that the working group's processes and methods reflected three key principles. First, the group used consensus-based decision making. Second, the group relied largely on existing academic literature and models, including technical assistance from outside resources. Third, the group took steps to disclose limitations and incorporate new information by considering public comments and revising the estimates as updated research became available.

Used Consensus-Based Decision Making

All of the participants we spoke with said that the working group used a consensus-based approach for making key decisions on developing the social cost of carbon estimates. Most participants said that the working group's overall approach was open and collegial, and that participants had many opportunities to make contributions and raise issues for discussion that were important to them.

OMB staff stated that the working group did not assign roles or responsibilities, and many participants told us that different working group participants and agencies volunteered to take responsibility for various aspects of the development of the estimates that fell within their particular areas of expertise. For example, OMB staff stated that, while OMB and the Council of Economic Advisers were the official leaders of the working group meetings, all EOP offices that participated played a large role during the meetings, and discussions were informal. According to these staff and other officials we spoke with, participants could generally choose the extent of their involvement, and all participants' contributions were considered equally.

According to many participants, the Council of Economic Advisers coordinated drafting the Technical Support Document, including gathering feedback from working group members. Specifically, they told us that, following the meetings, officials from the Council of Economic Advisers summarized the group discussions to include in the latest draft of the Technical Support Document and circulated draft sections of the Technical Support Document for the working group to review. For example, a participant told us that he raised concerns about whether the Technical Support Document provided adequate information on domestic measures of the social cost of carbon. The participant said that, in response to this feedback, the working group decided to include a separate discussion in the Technical Support Document on estimating domestic benefits and costs. The Technical Support Document states that reported domestic effects should be calculated using a range of values from 7 to 23 percent of the global measure of the social cost of carbon, although it cautions that these values are approximate, provisional, and highly speculative due to limited evidence. None of the participants we spoke with expressed concerns about how their contributions were incorporated into the final Technical Support Document. The participants generally stated that they were satisfied that the final Technical Support Document successfully addressed individual comments on the draft version and the overall consensus of the working group and its participating offices and agencies.

Relied on Existing Academic Literature and Models

The Technical Support Document states that the main objective of the working group was to develop a range of estimates of the social cost of carbon using a defensible set of modeling inputs based on existing academic literature. Many participants confirmed that the working group relied largely on existing academic literature and models to develop its estimates. According to the Technical Support Document and many participants we spoke with, the working group calculated its estimates using three models that integrate climate and economic data into a single modeling framework for estimating future economic effects resulting from climate change.[24] In general, each model translates carbon dioxide emissions scenarios into changes in greenhouse gas concentrations in the atmosphere, greenhouse gas concentrations in the atmosphere into temperature changes, and temperature changes into net economic effects (i.e., damages and benefits). However, each model uses its own methods to estimate these effects. The Technical Support Document states that the three models are frequently cited in peer-reviewed literature. They have also been used in climate assessments by the Intergovernmental Panel on Climate Change—an organization within the United Nations that assesses scientific, technical, and economic information on the effects of climate change. In addition, the National Research Council of the National Academies recognized these three models as three of the most widely used models of their kind.[25]

Many participants told us that the working group spent most of its meeting time reviewing and discussing academic literature to help decide on values for three key modeling inputs to run in each model. The key modeling inputs the working group selected were based on data from prevalent research organizations, such as the Stanford Energy Modeling Forum, and reflected the wide uncertainty in the academic literature, according to the Technical Support Document.[26] These inputs were as follows:

- scenarios for future population and economic growth (i.e., gross domestic product) and carbon dioxide emissions,
- a measure of the climate's responsiveness to increased concentrations of greenhouse gases in the atmosphere—known as equilibrium climate sensitivity,[27] and
- discount rates.

Several participants told us that different meetings focused on different modeling inputs and included technical presentations by participants with expertise in each technical area. For example, due to their previous experience working with the models, EPA officials made presentations on how each model works. OMB staff stated that the technical presentations focused on the academic materials cited in the Technical Support Document, including dozens of peer-reviewed journal articles. They also said that all technical decisions discussed in the Technical Support Document were arrived at by consensus through this process. Several participants said that a significant amount of the group's discussions focused on selecting discount rates that best reflect the most current academic literature, while also comporting with OMB's guidance in Circular A-4. The Technical Support Document cites guidance from Circular A-4 in its discussion of many technical topics, including its selection of discount rates. It states that the discount rate (i.e., 3 percent) used to calculate the central value of the social cost of carbon estimates is consistent with Circular A-4 guidance. Some working group participants told us that they recognized the importance of using OMB guidance, including Circular A-4, in developing the Technical Support Document. The Technical Support Document states that the working group decided to calculate estimates for several discount rates (2.5, 3, and 5 percent) because the academic literature shows that the social cost of carbon is highly sensitive to the discount rate chosen, and because no consensus exists on the appropriate rate. It further states that, in light of such uncertainties, the working group determined that these three discount rates reflect reasonable judgments about the appropriate rate to use. Several participants stated that the working group chose this approach to capture varied concerns and interests, including participants' respective knowledge of the academic literature, on selecting the discount rate.

Once the working group agreed on these modeling inputs, EPA officials supervised their use in running the models to calculate the social cost of carbon estimates. All other model assumptions and features were unchanged by the working group, which weighted each model equally to calculate the final estimates. Several participants stated that an important principle for the leaders of the working group was that the working group reach consensus on the modeling inputs before running the models and agree, in advance, to accept the results based on the inputs selected, whatever the outcome. Through this approach, the working group developed a set of four social cost of carbon estimates for use in regulatory impact analyses. The first three values are based on the average of the estimates produced by all three models and selected modeling inputs at the three discount rates chosen. The fourth value was included to represent higher-than-expected economic impacts from climate change, and it is based on an average of certain values produced by each model at a 3 percent discount rate.[28] To capture uncertainties involved in regulatory impact analysis, the Technical Support Document emphasizes the importance of agencies considering all four estimates when conducting analyses.

According to EPA documentation and several participants, groups from outside the federal government did not participate in the working group, but the working group used some outside resources, specifically technical assistance. As noted in the Technical Support Document, the working group explored technical literature in relevant fields for developing the social cost of carbon estimates. Members of the working group sometimes contacted researchers or developers of key data in an effort to ensure that the working group had a clear understanding of the information and how to use it. For example, according to several participants, members of the working group consulted with lead authors of a chapter on climate sensitivity that appears in the *Fourth Assessment Report of the Intergovernmental Panel on Climate Change*.[29] According to the Technical Support Document, after consulting with the chapter authors, the working group was able to make some decisions to assist with statistical analyses needed to develop the social cost of carbon estimates. Many participants stated that the working group also consulted with the developers of the models used by the group to develop the estimates. For example, EPA officials told us that, while they conducted runs for one model that was readily available to the public, they spent a few days training with the developer of a second model before using it to conduct runs. They also contracted with the developer of a third model to run the model according to the decisions reached by the working group. They stated that they ran all of the 2013 estimates themselves, but that they continued to consult with the model developers to do so.

According to many participants and the 2013 update to the Technical Support Document, the only changes made to the models used for the 2013 revisions were those that the model developers incorporated into the latest versions of the models and that were subsequently used in peer-reviewed academic literature. Specifically, the developers updated the academic models to reflect new scientific information, such as in sea level rise and associated damages, resulting in higher estimates.[30] The working group did not make changes in the modeling inputs that it used for the 2010 estimates.[31] Several participants said that, while the original working group included frequent, hours-long meetings over several months, the working group assembled to discuss the 2013 revisions only met a few times. According to the 2010 Technical Support Document, the working group is committed to updating its estimates as the science and economic understanding of climate change and its impacts on society improve over time.

Took Steps to Disclose Limitations and Incorporate New Information

According to several participants and the Technical Support Document, the working group's processes and methods took steps to disclose limitations and incorporate new information by considering public comments and revising the estimates as updated economic and scientific research became available. The Technical Support Document discusses several limitations of its estimates and areas that the working group identified as being in particular need of additional exploration and research. For example, it points out that none of the three models accounts for damages from wildlife loss or ocean acidification caused by carbon dioxide emissions. Also, the models cannot completely predict how technology may adapt to warmer temperatures. In addition, according to the Technical Support Document, the models may not fully consider the effects of damages due to potential catastrophic events, such as the melting of Antarctic ice sheets. As a result of such limitations, the models may underestimate damages from increased carbon emissions, according to the Technical Support Document. The Technical Support Document states that, as a result of these limitations, the social cost of carbon estimates should continue to evolve as knowledge is gained, and available models improve. Some of the participating agencies have incorporated discussions of these limitations into regulatory impact analyses using social cost of carbon estimates. For example, in a 2012 rule setting pollution standards for certain power plants, EPA noted that the social cost of carbon estimates are subject to limitations and uncertainties.[32]

Over the years, there have been opportunities for public comment on the various individually developed and working group estimates of the social cost of carbon for regulatory impact analysis, and several participants stated that these estimates were developed with input from the public. Since 2008, agencies have published over three dozen regulatory actions for public comment in the *Federal Register* that use various social cost of carbon estimates in regulatory impact analyses. While some of them specifically sought comments on the development of the social cost of carbon estimates used, and others did not, these regulatory actions were open to public comment, in general, for approximately 60 days and, according to OMB staff and other participants, agencies received many comments on the estimates through this process. Several participants stated that, while they discussed such public comments during working group meetings, individual agencies typically do not coordinate formally with other agencies on their reviews of comments received. According to the Technical Support Document, the working group convened, in part, to consider public comments on issues related to the social cost of carbon. After considering public comments on the interim values that agencies used in several rules, the working group developed the Technical Support Document, according to these participants and to the Technical Support Document. Several participants told us that the working group decided to revise the estimates for the first time in 2013 after agencies received a number of public comments encouraging revisions because the models used to develop the 2010 estimates had been subsequently updated and used in peer-reviewed academic literature. OMB staff stated that this theme was reflected in several public comments on regulations using the 2010 estimates.

In November 2013, OMB published a request in the *Federal Register* for public comments on all aspects of the Technical Support Document and its use of the models to develop estimates of the social cost of carbon.[33] The notice stated that OMB is particularly interested in comments on the selection of the models for use in developing the estimates, how the distribution of estimates should be represented in regulatory impact analyses, and the strengths and limitations of the overall approach. OMB staff told us that they decided to issue the request in response to calls for additional transparency, and that they received over 100 unique comments and thousands of identical form-letter comments in response to the request. They said that, since they were still reviewing the comments received, they had not yet decided on steps for responding to them, but that they expect to review them with the working group to determine whether they could inform future updates to the Technical Support Document. OMB staff stated that they have already made most of the comments publicly available online at http://www.regulations.gov/ and that all of the comments would be made available soon.

The Technical Support Document states that the working group would regularly revisit the social cost of carbon estimates as new information becomes available due to improved scientific and economic research. The Technical Support Document set a goal of revisiting the estimates within 2 years, or when substantially updated models become available. Many participants told us that, to revise the estimates in 2013, the working group met only a few times and mostly for participants from EPA to present information about updates made to the models since the group last met in 2010. The updates touched on a variety of issues, including how some models represent damages from sea level rise. The 2013 update to the Technical Support Document states that it acknowledges the continued limitations described in the original Technical Support Document, and that it updates the estimates based on new versions of the underlying models without revisiting the working group's decisions on modeling inputs. Several participants stated that they reviewed drafts of the 2013 update to the Technical Support Document, but that there was little new information to review because only the models had been updated. In addition to stating that the working group would regularly revisit its estimates, the Technical Support Document states that the working group will continue to support research to improve the estimates and hopes to develop methods to value other greenhouse gases as part of its ongoing work.[34]

Agency Comments

We provided a draft of this report for review and comment to the Departments of Agriculture, Commerce, Energy, Transportation, and the Treasury; EPA; and OMB. In its written comments, the Department of the Treasury stated that the draft report does a good job of capturing the interagency process through which the estimates of the social cost of carbon were developed. In oral comments provided on July 15, 2014, OMB staff confirmed that OMB generally agreed with the report findings. OMB staff also provided technical comments, which we incorporated into the report, as appropriate. The Department of Energy and EPA provided technical comments only, which we incorporated into the report, as appropriate. In e-mails received on July 1, July 9, and July 14, 2014, respectively, the liaisons from the Departments of Agriculture, Commerce, and Transportation stated that the departments did not have any comments on the draft report.

J. Alfredo Gómez
Director,
Natural Resources and Environment

Appendix I: Regulatory Actions, by Agency and Type of Social Cost of Carbon Estimates Used, 2008-2014

This appendix lists regulatory actions from 2008 to 2014 and the type of social cost of carbon estimates used (i.e., individually developed, interim, 2010, or 2013) in the actions' regulatory impact analyses. For each regulatory action, table 3 lists the date published in the *Federal Register*, the agency conducting the action, the name and status of the rule associated with the action, and the action's citation in the *Federal Register*.

Table 3. Regulatory Actions, by Agency and Type of Social Cost of Carbon Estimates Used in the Regulatory Impact Analysis, 2008-2014

Date published in the *Federal Register*	Agency	Rule	Status of rule	*Federal Register*citation
Individually developed agency estimates				
May 2, 2008	Department of Transportation (Transportation), National Highway Traffic Safety Administration (NHTSA)	Average Fuel Economy Standards, Passenger Cars and Light Trucks; Model Years 2011-2015	Proposed	73 Fed. Reg.24,352
July 30, 2008	Environmental Protection Agency (EPA)	Regulating Greenhouse Gas Emissions Under the Clean Air Act	Advanced Notice of Proposed Rulemaking	73 Fed. Reg.44,354
Aug. 25, 2008	Department of Energy (Energy)	Energy Conservation Program for Commercial and Industrial Equipment: Energy Conservation Standards for Commercial Ice-Cream Freezers; Self-Contained Commercial Refrigerators, Commercial Freezers, and Commercial Refrigerator-Freezers Without Doors; and Remote Condensing Commercial Refrigerators, Commercial Freezers, and Commercial Refrigerator-Freezers	Proposed	73 Fed. Reg.50,072
Oct. 7, 2008	Energy	Energy Conservation Program for Commercial and Industrial Equipment: Packaged Terminal Air Conditioner and Packaged Terminal Heat Pump Energy Conservation Standards	Final	73 Fed. Reg.58,772
Oct. 17, 2008	Energy	Energy Conservation Program: Energy Conservation Standards for Certain Consumer Products (Dishwashers, Dehumidifiers, Electric and Gas Kitchen Ranges and Ovens, and Microwave Ovens) and for Certain Commercial and Industrial Equipment (Commercial Clothes Washers)	Proposed	73 Fed. Reg.62,034

Table 3. (Continued)

Date published in the *Federal Register*	Agency	Rule	Status of rule	*Federal Register*citation
Jan. 9, 2009	Energy	Energy Conservation Program for Commercial and Industrial Equipment: Energy Conservation Standards for Commercial Ice-Cream Freezers; Self-Contained Commercial Refrigerators, Commercial Freezers, and Commercial Refrigerator-Freezers Without Doors; and Remote Condensing Commercial Refrigerators, Commercial Freezers, and Commercial Refrigerator-Freezers	Final	74 Fed. Reg.1092
Mar. 30, 2009	Transportation, NHTSA	Average Fuel Economy Standards Passenger Cars and Light Trucks Model Year 2011	Final	74 Fed. Reg.14,196
Apr. 8, 2009	Energy	Energy Conservation Program: Energy Conservation Standards for Certain Consumer Products (Dishwashers, Dehumidifiers, Microwave Ovens, and Electric and Gas Kitchen Ranges and Ovens) and for Certain Commercial and Industrial Equipment (Commercial Clothes Washers)	Final	74 Fed. Reg.16,040
Apr. 13, 2009	Energy	Energy Conservation Program: Energy Conservation Standards for General Service Fluorescent Lamps and Incandescent Reflector Lamps	Proposed	74 Fed. Reg.16,920
May 26, 2009	EPA	Regulation of Fuels and Fuel Additives: Changes to Renewable Fuel Standard Program	Proposed	74 Fed. Reg.24,904
May 29, 2009	Energy	Energy Conservation Program: Energy Conservation Standards for Refrigerated Bottled or Canned Beverage Vending Machines	Proposed	74 Fed. Reg.26,020
Mar. 26, 2010	EPA	Regulation of Fuels and Fuel Additives: Changes to Renewable Fuel Standard Program	Final	75 Fed. Reg.14,670
June 21, 2010	EPA	Hazardous and Solid Waste Management System; Identification and Listing of Special Wastes; Disposal of Coal Combustion Residuals From Electric Utilities	Proposed	75 Fed. Reg.35,128
2010 governmentwide estimates				

Date published in the *Federal Register*	Agency	Rule	Status of rule	*Federal Register*citation
Mar. 9, 2010	Energy	Energy Conservation Program: Energy Conservation Standards for Small Electric Motors	Final	75 Fed. Reg.10,874
Apr. 16, 2010	Energy	Energy Conservation Program: Energy Conservation Standards for Residential Water Heaters, Direct Heating Equipment, and Pool Heaters	Final	75 Fed. Reg.20,112
May 7, 2010	EPA and Transportation, NHTSA	Light Duty Vehicle Greenhouse Gas Emission Standards and Corporate Average Fuel Economy Standards	Final	75 Fed. Reg.25,324
May 28, 2010	Transportation, Federal Aviation Administration	Automatic Dependent Surveillance—Broadcast (ADS-B) Out Performance Requirements to Support Air Traffic Control (ATC) Service	Final	75 Fed. Reg.30,160
Aug. 2, 2010	EPA	Federal Implementation Plans To Reduce Interstate Transport of Fine Particulate Matter and Ozone	Proposed	75 Fed. Reg.45,210
Sep. 9, 2010	EPA	National Emission Standards for Hazardous Air Pollutants from the Portland Cement Manufacturing Industry and Standards of Performance for Portland Cement Plants	Final	75 Fed. Reg.54,970
Oct. 14, 2010	EPA	Standards of Performance for New Stationary Sources and Emission Guidelines for Existing Sources: Sewage Sludge Incineration Units	Proposed	75 Fed. Reg.63,260
Nov. 30, 2010	EPA and Transportation, NHTSA	Greenhouse Gas Emissions Standards and Fuel Efficiency Standards for Medium- and Heavy-Duty Engines and Vehicles	Proposed	75 Fed. Reg.74,152
Mar. 14, 2011	EPA	National Emission Standards for Hazardous Air Pollutants: Mercury Emissions from Mercury Cell Chlor-Alkali Plants	Supplementa l Proposed Rule	76 Fed. Reg.13,852
Mar. 21, 2011	EPA	Standards of Performance for New Stationary Sources and Emission Guidelines for Existing Sources: Sewage Sludge Incineration Units	Final	76 Fed. Reg.15,372
Mar. 21, 2011	EPA	National Emission Standards for Hazardous Air Pollutants for Major Sources: Industrial, Commercial, and Institutional Boilers and Process Heaters	Final	76 Fed. Reg.15,608

Table 3. (Continued)

Date published in the *Federal Register*	Agency	Rule	Status of rule	*Federal Register*citation
Apr. 11, 2011	Energy	Energy Conservation Program: Energy Conservation Standards for Fluorescent Lamp Ballasts	Proposed	76 Fed. Reg.20,090
Apr. 21, 2011	Energy	Energy Conservation Program: Energy Conservation Standards for Residential Clothes Dryers and Room Air Conditioners	Direct Final	76 Fed. Reg.22,454
June 27, 2011	Energy	Energy Conservation Program: Energy Conservation Standards for Residential Furnaces and Residential Central Air Conditioners and Heat Pumps	Direct Final	76 Fed. Reg.37,408
Aug. 8, 2011	EPA	Federal Implementation Plans: Interstate Transport of Fine Particulate Matter and Ozone and Correction of SIP Approvals[a]	Final	76 Fed. Reg.48,208
Sep. 15, 2011	EPA and Transportation, NHTSA	Greenhouse Gas Emissions Standards and Fuel Efficiency Standards for Medium- and Heavy-Duty Engines and Vehicles	Final	76 Fed. Reg.57,106
Sep. 15, 2011	Energy	Energy Conservation Program: Energy Conservation Standards for Residential Refrigerators, Refrigerator-Freezers, and Freezers	Final	76 Fed. Reg.57,516
Nov. 14, 2011	Energy	Energy Conservation Program: Energy Conservation Standards for Fluorescent Lamp Ballasts	Final	76 Fed. Reg.70,548
Dec. 1, 2011	EPA and Transportation, NHTSA	2017 and Later Model Year Light-Duty Vehicle Greenhouse Gas Emissions and Corporate Average Fuel Economy Standards	Proposed	76 Fed. Reg.74,854
Dec. 23, 2011	EPA	Commercial and Industrial Solid Waste Incineration Units: Reconsideration and Proposed Amendments; Non-Hazardous Secondary Materials That Are Solid Waste	Proposed	76 Fed. Reg.80,452
Jan. 17, 2012	Energy	Energy Conservation Program for Certain Industrial Equipment: Energy Conservation Standards and Test Procedures for Commercial Heating, Air-Conditioning, and Water-Heating Equipment	Proposed	77 Fed. Reg.2356
Feb. 10, 2012	Energy	Energy Conservation Program: Energy Conservation Standards for	Proposed	77 Fed. Reg.7282

Date published in the *Federal Register*	Agency	Rule	Status of rule	*Federal Register*citation
		Distribution Transformers		
Feb. 14, 2012	Energy	Energy Conservation Program: Energy Conservation Standards for Standby Mode and Off Mode for Microwaves	Supplemental Notice of Proposed Rulemaking	77 Fed. Reg.8526
Feb. 16, 2012	EPA	National Emission Standards for Hazardous Air Pollutants from Coal- and Oil-Fired Electric Utility Steam Generating Units and Standards of Performance for Fossil-Fuel-Fired Electric Utility, Industrial-Commercial-Institutional, and Small Industrial-Commercial-Institutional Steam Generating Units	Final	77 Fed. Reg.9304
Mar. 27, 2012	Energy	Energy Conservation Program: Energy Conservation Standards for Battery Chargers and External Power Supplies	Proposed	77 Fed. Reg.18,478
Apr. 13, 2012	EPA	Standards of Performance for Greenhouse Gas Emissions for New Stationary Sources: Electric Utility Generating Units	Proposed	77 Fed. Reg.22,392
May 30, 2012	Energy	Energy Conservation Program: Energy Conservation Standards for Residential Dishwashers	Direct Final	77 Fed. Reg.31,918
May 31, 2012	Energy	Energy Conservation Program: Energy Conservation Standards for Residential Clothes Washers	Direct Final	77 Fed. Reg.32,308
Oct. 15, 2012	EPA and Transportation, NHTSA	2017 and Later Model Year Light-Duty Vehicle Greenhouse Gas Emissions and Corporate Average Fuel Economy Standards	Final	77 Fed. Reg.62,624
Apr. 18, 2013	Energy	Energy Conservation Program: Energy Conservation Standards for Distribution Transformers	Final	78 Fed. Reg.23,336
June 7, 2013	EPA	Effluent Limitations Guidelines and Standards for the Steam Electric Power Generating Point Source Category	Proposed	78 Fed. Reg.34,432
2013 revised governmentwide estimates				
June 17, 2013	Energy	Energy Conservation Program: Energy Conservation Standards for Standby Mode and Off Mode for Microwave Ovens	Final	78 Fed. Reg.36,316
Aug. 20, 2013	Energy	Energy Conservation Program: Energy Conservation Standards for Metal Halide Lamp Fixtures	Proposed	78 Fed. Reg.51,464

Table 3. (Continued)

Date published in the *Federal Register*	Agency	Rule	Status of rule	*Federal Register*citation
Sep. 11, 2013	Energy	Energy Conservation Program: Energy Conservation Standards for Walk-In Coolers and Freezers	Proposed	78 Fed. Reg.55,782
Sep. 11, 2013	Energy	Energy Conservation Program: Energy Conservation Standards for Commercial Refrigeration Equipment	Proposed	78 Fed. Reg.55,890
Oct. 25, 2013	Energy	Energy Conservation Program for Consumer Products: Energy Conservation Standards for Residential Furnace Fans	Proposed	78 Fed. Reg.64,068
Dec. 6, 2013	Energy	Energy Conservation Program: Energy Conservation Standards for Commercial and Industrial Electric Motors	Proposed	78 Fed. Reg.73,590
Jan. 8, 2014	EPA	Standards of Performance for Greenhouse Gas Emissions from New Stationary Sources: Electric Utility Generating Units	Proposed	79 Fed. Reg.1430
Feb. 10, 2014	Energy	Energy Conservation Program: Energy Conservation Standards for Metal Halide Lamp Fixtures	Final	79 Fed. Reg.7746
Feb. 10, 2014	Energy	Energy Conservation Program: Energy Conservation Standards for External Power Supplies	Final	79 Fed. Reg.7846
Mar. 4, 2014	Energy	Energy Conservation Program: Energy Conservation Standards for Commercial Clothes Washers	Proposed	79 Fed. Reg.12,302
Mar. 17, 2014	Energy	Energy Conservation Program: Energy Conservation Standards for Automatic Commercial Ice Makers	Proposed	79 Fed. Reg.14,846
Mar. 28, 2014	Energy	Energy Conservation Program: Energy Conservation Standards for Commercial Refrigeration Equipment	Final	79 Fed. Reg.17,726
Apr. 29, 2014	Energy	Energy Conservation Program: Energy Conservation Standards for General Service Fluorescent Lamps and Incandescent Reflector Lamps	Proposed	79 Fed. Reg.24,068
May 29, 2014	Energy	Energy Conservation Program: Energy Conservation Standards for Commercial and Industrial Electric Motors	Final	79 Fed. Reg.30,934
June 3, 2014	Energy	Energy Conservation Program: Energy Conservation Standards for Walk-In Coolers and Freezers	Final	79 Fed. Reg.32,050
June 18, 2014	EPA	Carbon Pollution Emission	Proposed	79 Fed.

Date published in the *Federal Register*	Agency	Rule	Status of rule	*Federal Register* citation
		Guidelines for Existing Stationary Sources: Electric Utility Generating Units		Reg.34,830

Sources: Environmental Protection Agency and Federal Register. | GAO-14-663.

Notes:

Regulatory actions in this table are as of June 18, 2014.

In 2008 and early 2009, individual estimates of the social cost of carbon were developed by each agency and typically based on estimates published in academic literature. The interim governmentwide estimates were developed in early 2009 by the Interagency Working Group on Social Cost of Carbon and derived from an average of selected estimates published in academic literature. The 2010 governmentwide estimates were developed by the Interagency Working Group on Social Cost of Carbon and issued in its February 2010 Technical Support Document. The 2013 revised governmentwide estimates were developed by the Interagency Working Group on Social Cost of Carbon and issued in a May 2013 update to the Technical Support Document, which was reissued with minor technical corrections in November 2013.

[a] SIP refers to State Implementation Plan.

End Notes

[1] The social cost of carbon (measured in dollars per metric ton of carbon dioxide) is the monetized net effects (damages and benefits) associated with an incremental increase in carbon emissions in a given year. Estimates of the social cost of carbon depend on the data and the models used to calculate them and can include a wide range of damage categories, such as projected changes in net agricultural productivity, human health, and property damages from increased flood risk due to increased carbon emissions. Monetization is the process of estimating the dollar value of benefits and costs.

[2] Carbon dioxide is a greenhouse gas recognized as a major contributor to climate change. Concentrations of greenhouse gases—including carbon dioxide, methane, nitrous oxide, and synthetic chemicals such as fluorinated gases—trap heat in the atmosphere and prevent it from returning to space.

[3] Interagency Working Group on Social Cost of Carbon, United States Government, Technical Support Document: Social Cost of Carbon for Regulatory Impact Analysis under Executive Order 12866 (Washington, D.C.: February 2010).

[4] Interagency Working Group on Social Cost of Carbon, United States Government, Technical Support Document: Technical Update of the Social Cost of Carbon for Regulatory Impact Analysis under Executive Order 12866 (Washington, D.C.: May 2013). This document was reissued with minor technical corrections in November 2013.

[5] According to the Technical Support Document, the working group consisted of participants from the Council of Economic Advisers, Council on Environmental Quality, EPA, National Economic Council, Office of Energy and Climate Change, OMB, Office of Science and

Technology Policy, and the Departments of Agriculture, Commerce, Energy, Transportation, and the Treasury. In March 2011, the Office of Energy and Climate Change joined the Domestic Policy Council.

[6] Exec. Order No. 12866, 58 Fed. Reg. 51,735 (Sept. 30, 1993). Other significant regulatory actions include those that are likely to result in a rule that may create a serious inconsistency or otherwise interfere with an action taken or planned by another agency; materially alter the budgetary impact of entitlements, grants, user fees, or loan programs or the rights and obligations of recipients thereof; or raise novel legal or policy issues arising out of legal mandates, the President's priorities, or the principles set forth in Executive Order 12866.

[7] OMB, Circular A-4: Regulatory Analysis (Sept. 17, 2003).

[8] When the benefits and costs of a regulatory action will occur in the future, agencies must determine the present value of future benefits and costs by applying an appropriate discount rate—the interest rate used to convert benefits and costs occurring in different time periods to a common present value.

[9] Circular A-4 states that agencies should discount future benefits and costs using rates of 3 and 7 percent but notes that agencies may, in addition, consider a lower discount rate if a rule will have important intergenerational benefits or costs. In July 2014, we reported on the application of the guidance in Circular A-4 and the Technical Support Document and made recommendations to OMB to help clarify the relationship between those two documents. See GAO, Environmental Regulation: EPA Should Improve Adherence to Guidance for Selected Elements of Regulatory Impact Analyses, GAO-14-519 (Washington, D.C.: July 18, 2014).

[10] Exec. Order No. 13563, 76 Fed. Reg. 3821 (Jan. 18, 2011).

[11] Average Fuel Economy Standards for Light Trucks Model Years 2008-2011, 71 Fed. Reg. 17,566 (Apr. 6, 2006). According to EPA officials, other regulations at the time did not typically quantify changes in carbon emissions.

[12] Ctr. For Biological Diversity v. Nat'l Highway Traffic Safety Admin., 538 F.3d 1172, 1203 (9th Cir. 2008). The Ninth Circuit issued the 2008 opinion after vacating and withdrawing its prior opinion, 508 F.3d 508, issued on Nov. 15, 2007.

[13] Id. at 1200.

[14] The benefits and costs of reducing most greenhouse gas emissions, including carbon dioxide, differ from most other benefits and costs in at least two respects: (1) greenhouse gas emissions can contribute to global damages even when emitted in the United States because these emissions can disperse widely throughout the atmosphere, and (2) these emissions generally remain in the atmosphere for years, causing subsequent long-term damages. While Circular A-4 states that agencies should generally estimate domestic benefits and costs of regulations, it also provides latitude to include global economic effects resulting from regulations when relevant and states that such effects should be reported separately and in addition to domestic effects.

[15] Energy Conservation Program: Energy Conservation Standards for Refrigerated Bottled or Canned Beverage Vending Machines, 74 Fed. Reg. 44,914 (Aug. 31, 2009).

[16] The working group calculated five interim estimates of the social cost of carbon using different discount rate scenarios and referred to $19—the middle of the five estimates—as the "central value."

[17] Energy Conservation Program: Energy Conservation Standards for Small Electric Motors, 75 Fed. Reg. 10,874 (Mar. 9, 2010).

[18] Energy Conservation Program: Energy Conservation Standards for Standby Mode and Off Mode for Microwave Ovens, 78 Fed. Reg. 36,316 (June 17, 2013).

[19] We previously reported that four of these EOP offices—the Council on Environmental Quality, Office of Energy and Climate Change, OMB, and Office of Science and Technology Policy—provide high-level policy direction for federal climate change programs and activities and commonly lead formal and informal interagency initiatives on related issues. See GAO, Climate Change: Improvements Needed to Clarify National Priorities and Better Align Them with Federal Funding Decisions, GAO-11-317 (Washington, D.C.: May 20, 2011).

[20] The 2013 update to the Technical Support Document adds that Executive Order 13563, issued after the working group developed the 2010 social cost of carbon estimates, commits the administration to regulatory decision making based on the best available science.

[21] GAO, Standards for Internal Control in the Federal Government, GAO/AIMD-00-21.3.1 (Washington, D.C.: November 1999).

[22] GAO, GAO-11-317; Managing for Results: Key Considerations for Implementing Interagency Collaborative Mechanisms, GAO-12-1022 (Washington, D.C.: Sept. 27, 2012); and Managing for Results: Implementation Approaches Used to Enhance Collaboration in Interagency Groups, GAO-14-220 (Washington, D.C.: Feb. 14, 2014).

[23] GAO-12-1022.

[24] The three models are Dynamic Integrated Climate and Economy (DICE), Climate Framework for Uncertainty, Negotiation, and Distribution (FUND), and Policy Analysis of the Greenhouse Effect (PAGE). They were first developed in the early 1990s by researchers acknowledged as leaders in their field and are updated regularly based on new developments in climate and economic research.

[25] National Research Council, Hidden Costs of Energy: Unpriced Consequences of Energy Production and Use (Washington, D.C.: National Academies Press, 2010).

[26] The Stanford Energy Modeling Forum is an international forum for sharing and facilitating discussions on energy policy and global climate issues among researchers.

[27] Equilibrium climate sensitivity is the long-term increase in the annual global-average surface temperature from a sustained doubling of the concentration of carbon dioxide in the atmosphere relative to preindustrial levels of the concentration of carbon dioxide in the atmosphere.

[28] According to the Technical Support Document, the working group determined the fourth value by combining the values appearing at the furthest reaches of the distributions produced by each model. For this purpose, the working group used values produced from all three models for the 95th percentile at a 3 percent discount rate.

[29] Intergovernmental Panel on Climate Change, Climate Change 2007: The Physical Science Basis. Contribution of Working Group I to the Fourth Assessment Report of the Intergovernmental Panel on Climate Change (S. Solomon, et al. [eds.])(Cambridge, UK: Cambridge University Press, 2007).

[30] This new scientific information included an explicit representation of sea level rise and associated damages, updated climate change adaptation assumptions, and updated damage functions for agricultural impacts.

[31] In January 2014, a former coleader of the working group discussed some of the reasons behind this approach in a presentation before the annual meeting of the American Economic Association, a leading economic interest group. See Cass Sunstein, "On Not Revisiting Official Discount Rates: Institutional Inertia and the Social Cost of Carbon" (paper presented at the annual meeting of the American Economic Association, Philadelphia, PA, Jan. 3, 2014). In 2013, another former coleader of the working group published a paper detailing the working group's methodology. See Greenstone, Michael et al., "Developing a

Social Cost of Carbon for U.S. Regulatory Analysis: A Methodology and Interpretation," Review of Environmental Economics and Policy 7, no. 1: 23-46 (2013).

[32] National Emission Standards for Hazardous Air Pollutants From Coal- and Oil-Fired Electric Utility Steam Generating Units and Standards of Performance for Fossil-FuelFired Electric Utility, Industrial-Commercial-Institutional, and Small Industrial-CommercialInstitutional Steam Generating Units, 77 Fed. Reg. 9304 (Feb. 16, 2012).

[33] Technical Support Document: Technical Update of the Social Cost of Carbon for Regulatory Impact Analysis under Executive Order No. 12866, 78 Fed. Reg. 70,586 (Nov. 26, 2013). In January 2014, OMB extended the public comment period through February 26, 2014. See Technical Support Document: Technical Update of the Social Cost of Carbon for Regulatory Impact Analysis Under Executive Order No. 12866, 79 Fed. Reg. 4359 (Jan. 27, 2014).

[34] In late 2010 and early 2011, EPA and the Department of Energy sponsored two workshops on valuing climate change damages for regulatory analysis. The agencies reported that they sponsored the workshops to prepare for and inform future working group activities. See ICF International, Workshop Report: Improving the Assessment and Valuation of Climate Change Impacts for Policy and Regulatory Analysis – Part 1 (January 2011); summary of workshop sponsored by EPA and the Department of Energy and titled "Modeling Climate Change Impacts and Associated Economic Damages" (Washington, D.C.: Nov. 18-19, 2010) and Workshop Report: Improving the Assessment and Valuation of Climate Change Impacts for Policy and Regulatory Analysis – Part 2 (March 2011); summary of workshop sponsored by EPA and the Department of Energy and titled "Research on Climate Change Impacts and Associated Economic Damages" (Washington, D.C., Jan. 27-28, 2011).

INDEX

D

E

F

G

O

P

Q

R

T

U

V

W

X

Y